Charlie

THE DOG WHO Came In From THE WILD

Hubble *and* Hattie

Lisa Tenzin-Dolma
with a Foreword by Marc Bekoff

Hubble & Hattie

The Hubble & Hattie imprint was launched in 2009 and is named in memory of two very special Westie sisters owned by Veloce's proprietors. Since the first book, many more have been added to the list, all with the same underlying objective: to be of real benefit to the species they cover, at the same time promoting compassion, understanding and respect between all animals (including human ones!) All Hubble & Hattie publications offer ethical, high quality content and presentation, plus great value for money.

More books from Hubble & Hattie –

Among the Wolves: Memoirs of a wolf handler (Shelbourne)
Animal Grief: How animals mourn (Alderton)
Babies, kids and dogs – creating a safe and harmonious relationship (Fallon & Davenport)
Because this is our home ... the story of a cat's progress (Bowes)
Camper vans, ex-pats & Spanish Hounds: from road trip to rescue – the strays of Spain (Coates & Morris)
Cat Speak: recognising & understanding behaviour (Rauth-Widmann)
Charlie – The dog who came in from the wild (Tenzin-Dolma)
Clever dog! Life lessons from the world's most successful animal (O'Meara)
Complete Dog Massage Manual, The – Gentle Dog Care (Robertson)
Dieting with my dog: one busy life, two full figures ... and unconditional love (Frezon)
Dinner with Rover: delicious, nutritious meals for you and your dog to share (Paton-Ayre)
Dog Cookies: healthy, allergen-free treat recipes for your dog (Schöps)
Dog-friendly Gardening: creating a safe haven for you and your dog (Bush)
Dog Games – stimulating play to entertain your dog and you (Blenski)
Dog Relax – relaxed dogs, relaxed owners (Pilguj)
Dog Speak: recognising & understanding behaviour (Blenski)
Dogs on Wheels: travelling with your canine companion (Mort)
Emergency First Aid for dogs: at home and away Revised Edition (Bucksch)
Exercising your puppy: a gentle & natural approach – Gentle Dog Care (Robertson & Pope)
Fun and Games for Cats (Seidl)
Gods, ghosts, and black dogs – the fascinating folklore and mythology of dogs (Coren)
Gymnastricks: Targeted muscle training for dogs (Mayer)
Helping minds meet – skills for a better life with your dog (Zulch & Mills)
Home alone and happy – essential life skills for preventing separation anxiety in dogs and puppies (Mallatratt)
Know Your Dog – The guide to a beautiful relationship (Birmelin)

Life skills for puppies – laying the foundation of a loving, lasting relationship (Zuch & Mills)
Living with an Older Dog – Gentle Dog Care (Alderton & Hall)
Miaow! Cats really are nicer than people! (Moore)
My cat has arthritis – but lives life to the full! (Carrick)
My dog has arthritis – but lives life to the full! (Carrick)
My dog has cruciate ligament injury – but lives life to the full! (Haüsler & Friedrich)
My dog has epilepsy – but lives life to the full! (Carrick)
My dog has hip dysplasia – but lives life to the full! (Haüsler & Friedrich)
My dog is blind – but lives life to the full! (Horsky)
My dog is deaf – but lives life to the full! (Willms)
My Dog, my Friend: heart-warming tales of canine companionship from celebrities and other extraordinary people (Gordon)
No walks? No worries! Maintaining wellbeing in dogs on restricted exercise (Ryan & Zulch)
Partners – Everyday working dogs being heroes every day (Walton)
Smellorama – nose games for dogs (Theby)
Swim to recovery: canine hydrotherapy healing – Gentle Dog Care (Wong)
A tale of two horses – a passion for free will teaching (Gregory)
Tara – the terrier who sailed around the world (Forrester)
The Truth about Wolves and Dogs: dispelling the myths of dog training (Shelbourne)
Waggy Tails & Wheelchairs (Epp)
Walking the dog: motorway walks for drivers & dogs revised edition (Rees)
When man meets dog – what a difference a dog makes (Blazina)
Winston ... the dog who changed my life (Klute)
The quite very actual adventures of Worzel Wooface (Pickles)
You and Your Border Terrier – The Essential Guide (Alderton)
You and Your Cockapoo – The Essential Guide (Alderton)
Your dog and you – understanding the canine psyche (Garratt)

For post publication news, updates and amendments relating to this book please visit www.hubbleandhattie.com/extras/HH4784

www.hubbleandhattie.com

First published in August 2015 by Veloce Publishing Limited, Veloce House, Parkway Farm Business Park, Middle Farm Way, Poundbury, Dorchester, Dorset, DT1 3AR, England. Reprinted January 2016. Fax 01305 250479/email info@hubbleandhattie.com/web www.hubbleandhattie.com ISBN: 978-1-845847-84-5 UPC: 6-36847-04784-9 © Lisa Tenzin-Dolma & Veloce Publishing Ltd 2015, 2016. Readers with ideas for books about animals, or animal-related topics, are invited to write to the editorial director of Veloce Publishing at the above address. British Library Cataloguing in Publication Data – A catalogue record for this book is available from the British Library. Typesetting, design and page make-up all by Veloce Publishing Ltd on Apple Mac. Printed in India by Replika Press

contents

Acknowledgements

First and foremost I thank Charlie for putting his trust in me and for being such an extraordinary teacher and friend. Denisa Munteanu saved Charlie's life when she took him into her rescue in Romania, and has celebrated each stage of his progress with us via the internet. Thank you, Denisa, for all that you do for the many souls who are fortunate enough to come into your tender care! And the trustees at Blind Dog Rescue UK paved the way for Charlie's new life to begin; thank you.

Skye, my Lurcher, has made so much of Charlie's adjustment possible and is, as always, an immense inspiration – he truly is a king among dogs. My daughter, Amber Tenzin-Dolma, has opened her heart to every sad and scared dog who has come through our doors, and has added her gentle energy to the healing process in each of them. My sons, Ryan and Oliver Lipscombe, and Daniel and Liam Tenzin-Dolma, have joined in offering support during the sticky times, and in celebrating Charlie's progress, either in person or at the other end of the phone.

My fabulous literary agent and good friend, Samantha Curtis, believed strongly that this book needed to be written, and has given tremendous encouragement throughout. Thank you so much, Sam! And huge thanks to Jude Brooks at Hubble & Hattie for making Charlie's story available for others to read.

Many thanks to Dr Marc Bekoff, who generously sent me research papers when I was searching for scientific information about feral dogs, soon after Charlie came to us. These were invaluable in helping me to understand what Charlie's life would have been like in the wild, and provided a great many insights into how I could help Charlie to adjust to his new life. Thank you, also, Marc, for reading the manuscript and writing the beautiful foreword!

Team Charlie helped us in so many ways, giving practical and emotional support and being our cheerleaders: Gina Holmes, who brought Charlie to us and has been a good friend to my dogs and myself; my vets, Amelia Welham and Liz Hornsby, for medical care, and Amelia for her long friendship and her compassion for all the dogs who have lived with us; Theo Stewart, close friend and colleague, who's been there every step of the way; Sarah Fisher, whose knowledge is awe-inspiring; holistic vets Sarah Drawbridge and Paula Kunkos, who treated Charlie during the months when his behaviour was very challenging; wolf and dog experts Toni Shelbourne and Isla Fishburn, who provided many insights into the wild aspect of Charlie's nature. Lisa Dickinson and Sue Beech were always just a phone call away – thank you!

The snapshots of Charlie were taken by Amber and me, and, of course, we weren't planning on sharing them widely, so they're typical family photos! Photographer Kerry James visited when the text was completed and took many beautiful photos for this book. Huge

Skye, Lisa and Charlie.

thanks, Kerry – I'll treasure your pictures for years to come!

I owe thanks to too many friends to name everyone in this book. Big hugs to Annie and Bryan Rawlings; Michael Eastwood; Marius Von Brasch; Kac Young; Marlene Morris; Caroline Wilkinson, and Ruth Bartleet. Beverley Cuddy of *Dogs Today* magazine has been hugely supportive. Many other friends, in person and online, were very much there for us, and I feel blessed to have all of them in my life.

Foreword

By Marc Bekoff

Charlie: The dog who came in from the wild describes how Lisa Tenzin-Dolma helped Charlie, a Romanian feral dog, adjust to his new life in her home in England. This book is a moving love story, and a testimony to the effectiveness of compassionate, loving, and force-free methods. It describes the immense rewards gained by all involved when the many challenges posed by the introduction of an unsocialized adult dog into domestic life were surmounted. This process involved a great deal of commitment, dedication and perseverance, and Lisa describes how she worked to earn Charlie's trust while guiding him to feel safe in a terrifyingly unfamiliar world.

Because I have co-written papers on the feralization of domestic dogs, and the social biology of free-ranging dogs with my then graduate student, Tom Daniels, Lisa contacted

me soon after Charlie's arrival, when information about his background emerged. She felt that my research into feral dogs would provide insights into Charlie's previous life so that she could enable him to make as smooth a transition as possible into his new life. Until recently, no formal research had been conducted into the socialization of feral dogs, and currently, a study of a feral dog named Parsifal is taking place in Turin, Italy.

I met Parsifal, and, as with Charlie, his behaviour seemed so much more wolf-like than dog-like that, if his appearance had been more like that of a husky, I would have been convinced that his genetic make-up was that of a wolf.

The company we keep can transform our lives, and our relationships with the non-human animals we invite into our homes and hearts can be extraordinarily enriching to all involved. This is especially so when we make the effort to understand our canine companions, and look at the world from their perspective. My research into animal emotions and the development of communication and social relationships in canids has convinced me that the inner lives of non-human animals are deeply rich and complex.

Cognitive ethology (the study of animal minds) has shown that non-human animals experience a richly varied range of emotions. We are only just beginning to delve beneath the surface, and each new discovery reveals that there are yet more hidden depths to explore. I am sure that future research will reveal not only interesting differences among species but also striking similarities across mammals, including ourselves.

Charlie had no previous experience of a relationship with humans. His developing bond with Lisa and her daughter carries a message for us: even a "wild soul," as Lisa calls him, is capable of forming powerful bonds and learning to adapt within an environment that is completely alien to him. If we each view ourselves as mirrors, both reflecting and being reflected in our relationships with each other and with other species, we can learn a great deal about ourselves, as well as about our non-human companions. I feel certain you will agree that this is among the very important take-home messages of *Charlie: The dog who came in from the wild*.

Marc Bekoff
Boulder, Colorado

Marc Bekoff

Marc Bekoff is Professor Emeritus of Ecology and Evolutionary Biology at the University of Colorado, Boulder, and a former Guggenheim Fellow. In 2000 he was awarded the Exemplar Award from the Animal Behavior Society for major long-term contributions to the field of animal behavior, and in 2009 he was presented with the Saint Francis of Assisi Award by the Auckland (New Zealand) SPCA.

Marc has published more than 1000 scientific and popular essays, and thirty books including *Minding Animals, The Ten Trusts* (with Jane Goodall), *The Emotional Lives of Animals, Animals Matter, Animals at Play: Rules of the Game, Wild Justice: The Moral Lives of Animals* (with Jessica Pierce), *The Animal Manifesto: Six Reasons for Expanding Our Compassion Footprint, Ignoring Nature No More: The Case For Compassionate Conservation, Jasper's Story: Saving Moon Bears, Why Dogs Hump and Bees Get Depressed: The Fascinating Science of Animal Intelligence, Emotions, Friendship, and Conservation, Rewilding Our Hearts: Building Pathways of Compassion and Coexistence*, and two editions of the *Encyclopedia of Animal Rights and Animal Welfare*, the *Encyclopedia of Animal Behavior*, and the *Encyclopedia of Human-Animal Relationships*.

In 2005 Marc was presented with The Bank One Faculty Community Service Award for the work he has done with children, senior citizens, and prisoners as part of Jane Goodall's Roots & Shoots programme.

His websites are marcbekoff.com and, with Jane Goodall, www.ethologicalethics.org.

one THE WILD SELF

My ears are ringing. Charlie, our one-eyed Romanian feral dog, has heard something in the distance – perhaps another dog barking, a car engine revving, or the red-tailed hawks calling as they fly high above us. He throws back his head to fully expose his throat, purses his lips into an 'O' shape and emits a lengthy, high decibel howl. Charlie is announcing to the world that he is here ... and that he wishes to have his presence acknowledged.

Charlie's howls are, fortunately, a source of deep fascination for my friends in the small village where we live. However, they're loud, and they carry across great distances. It's hard to ignore them, and they can be a strain on the eardrums if they continue for long. So, in the interests of maintaining harmonious neighbourly relations, I call Charlie indoors and reward him with praise and a liver treat for coming to me. He sits and offers me a paw in a gesture of trust and friendship, and I take it in my hand and hold it gently as I thank him.

Less than eighteen months previously, this feisty strength of character would have been hard to imagine, yet it was the reason for Charlie's survival in the wild during the first years of his life. When he first came to live with us he was in a state of paralysed shock and terror. He was totally unsocialized, and had never experienced a close relationship with a human, or been inside a home. In many ways, Charlie seemed more like a wolf than a dog, and he has travelled great distances through geographical location and newly-learned behavioural responses in the comparative blink of his single eye. I have learned a great deal from every dog I have lived and worked with – especially from Skye, my wise Deerhound/Greyhound/Saluki mix Lurcher, who has mentored all of my fostered and adopted dogs over the past seven years – but Charlie has been, without doubt, my greatest teacher.

The lessons that Charlie and I have absorbed from each other have relevance for each of us. We all experience the intense emotions that Charlie so eloquently expresses. We all have to find our way through the painful, thorny patches that occasionally block our way, and at times we all experience an urge

---

Nature's redness of tooth and claw is for real

---

to throw back our heads and sing out to make our presence known. This book tells the story of life with Charlie during his first eighteen months with us, but it is not the whole story because the events of his wild past can only be guessed at. Instead, what follows are chapters in the life of a dog whose world was turned upside down, and then gradually turned around again to find a different level.

Our beautiful dog's path to healing is also one that we all experience in our own individual ways after facing up to the challenges that life delivers to us. Charlie's tale prompted me to think about the true nature of humanity and dogdom, and a crucial element of our relationship involved striving to understand the conflict within Charlie between his wild nature and his dog nature.

WILD REALITY

Life in the wild, so symbolic of freedom, autonomy and a profound connection with the natural world, tends to be romanticized; yet the yearning to be connected with the world around us runs deep. Nature's redness of tooth and claw is for real in the world of a feral dog: daily life involves danger and a constant battle to stay alive, and necessitates being aware of every shadow that may lurk, ready to pounce from behind the next rock or tree. It's about finding the will, strength and courage to survive in lean times; about living in the moment and celebrating through play; about appreciating and deciphering the scents drifting on the breeze, and following the ones that tantalise.

The wild self still lives on in all of us, softened and tamed though we are on the surface, nowadays. It prompts us to listen to the messages our brains and bodies send us: to follow our noses, guts and intuition, and act instantly on those. The wild self is governed by self-preservation, by joy and spontaneity, and it pays close attention to the rushes of fear that keep us safe from potential harm. It whispers to us to take only what we need, and to leave tracks through life's thickets for those of our own kind to follow.

We humans are a domesticated species, who, like all animals, seek safety, comfort and happiness. Many of us find ourselves caught in the trappings of modern life, quarantined from nature by tall buildings and electric lights, coaxed to acquire things we don't need in order to feel good about ourselves; striving to fill an inner emptiness that hollows out when our roots dry

---

The wild self still lives on in all of us

---

up and shrivel through the loss of connection with our essential selves. We find solace in our companion animals – those missing links with nature that we bring into our homes. Frequently, in doing so, we enforce our will on them to tame and shape them as far from their instinctual nature as possible; yet our hearts, like theirs, beat to the rhythm of Nature's drum.

Sometimes our wild selves visit us in dreams, or through art, music, or dance, or making love. In my case, the arrival of a frightened feral dog on a cold, dark February evening in 2013 made me consciously aware of how powerful that primal nature is; of how the savage breast can, perhaps, be soothed, but will always retain its secrets, and will never truly be subsumed – nor should it be. Back then, I knew there were challenges ahead, but I had no idea that Charlie would break my heart numerous times, mend it just as often, turn my world upside down, and show me a new perspective.

As a small child I was intrigued by the foundation story of Rome: the tale of the semi-divine twin brothers, Romulus and Remus, who were born in Alba Longa. I read it over and over, and it left a deep impression. These are the bare bones of the story –

Taken from their mother at birth by their wicked great-uncle, Amulius, and abandoned to die by the river Tiber, the brothers were suckled by a she-wolf, and eventually fostered by a shepherd and his wife. In time, they learned of their true origins, gathered support, and overthrew

Charlie

Amulius, restored Numitor, their grandfather, to his rightful place on the throne, and decided to found a new city of their own. The twins each chose a location for this, but fought over which would be their new domain. Romulus killed Remus, built the city on his chosen site, and named it Rome.

The element of this story that captured my imagination was the she-wolf who sustained – instead of taking – the lives of two helpless babies. To me, she embodied true power – the elemental clarion call of the wild self who only devours what is necessary for survival, and whose role it is to nurture the seeds of the future.

Perhaps this echo of a childhood fascination was a touchstone for the extraordinary bond that emerged between Charlie and me. He was wild: untouched by the trappings of domestication and the need to conform to others' expectations. Charlie clearly belonged only to himself, and his friendship was (and is) a gift; never a right.

What follows is the story of Charlie's first eighteen months with us. His journey towards learning to live in the modern world that we take for granted has been a revelation for me, as well as for him, as I watched him, at first, struggle to accept his new life, and finally embrace it with open paws. Charlie is the living, breathing embodiment of the wild nature that inhabits each of us, and breaks forth occasionally in spontaneous wolf-howls, sudden ferocity, the capacity for boundless love, and dances of glee. It may be hidden much deeper within us but it is there, all the same, just waiting to be rediscovered.

Socialising Charlie was never meant to suppress his wildness, his essential self, and (thank goodness!), it hasn't. Teaching him to live comfortably and safely in our world has only been one part of our relationship. The privilege of becoming part of a strongly woven bond of love that encompasses Charlie, myself, and my lurcher, Skye, is the essence of our lives together. Learning to look at the world through Charlie's pale amber eye has changed my life forever.

VISIT HUBBLE AND HATTIE ON THE WEB: WWW.HUBBLEANDHATTIE.COM
WWW.HUBBLEANDHATTIE.BLOGSPOT.CO.UK
• DETAILS OF ALL BOOKS • SPECIAL OFFERS • NEWSLETTER • NEW BOOK NEWS

TWO Arrival

Charlie greets me at the door, his tail wagging madly. He wriggles past Skye and winds himself around my legs. I kneel down and he rubs himself ecstatically against me, spreading his scent across as wide an area as possible to stake his claim on me yet again. At the same time he burrows his nose wherever he can reach, gathering olfactory information about where I've been and with whom.

When I have been with other dogs, this investigatory process takes some time. Skye waits close by, patient as always, until Charlie has worn out his welcome and there's space enough for him to step into the gap and lick my cheek. Charlie leans against me, tongue lolling, his single, amber-coloured eye soft and squinty. His family group is reunited and all is now right in his world.

The daily occurrence of this scene fills me with joy. We often hear how things are 'meant to be,' and Charlie's presence in my life came about through a spur-of-the-moment impulse when I saw an urgent appeal from Blind Dog Rescue UK for foster carers for a group of dogs who were travelling to the UK from Romania. My elderly Greyhound, Duke, had recently passed away, and I was ready to foster again, so I emailed to say I would be happy to take in whichever dog most needed help. Charlie had received no other foster offers, so was allocated to me.

I had no idea back then that he was an unsocialized feral dog, and neither did the rescue, but as it was clear from the start that he had experienced no close contact with humans, I contacted Denisa Munteanu, his rescuer in Romania, to ask about his background. Denisa informed me that Charlie and his constant companion, Lenny, had first been sighted in May 2011. She had been tracking the pair for some time, leaving food when she could. She told me that they lived out in a field and were "very shy" of humans, and that she had taken them in when it was clear that Charlie needed surgery

---

Plus, we'd already fallen in love with him

---

to remove his badly injured left eye. Capturing him in December 2012, once she had tempted him close with food, was easier than with most free-ranging dogs. Whereas many dogs react aggressively to attempts at restraining them, Charlie crouched close to the ground, frozen with fear, which is still his primary response when he's very scared, though, nowadays, it's a very rare occurrence, fortunately.

Charlie was meant to be a short-term guest, staying only until a forever home could be found for him, but right from the beginning it was clear that he needed specialized care. Plus, we'd already fallen in love with him. Two weeks later my daughter, Amber, and I adopted him.

As a canine behaviour practitioner, I

Charlie

specialize in rescue dogs, and very fearful dogs who have undergone trauma and lost all trust in humans – yet Charlie was to be the greatest canine challenge of my life so far. Rebuilding the trust of a previously abused dog takes time and patience, and I've always loved to work with these sad souls, and watch them gradually blossom. But Charlie was unlike any other dog I had ever met, and he has taught me a great deal about connecting with the wild self; about patience, dealing with fear (my own and that of others, as well as Charlie's many fears), and about the sweet purity of unconditional love that transcends species, and ultimately connects us at core level.

FIRST STEPS TO HEALING

Creating bonds of trust with Charlie involved monitoring myself, as well as him, very closely at all times. He existed in a continual state of sheer terror: everything in the alien environment he was thrust into was strange and scary, and his past life in the wild had given him no coping mechanisms to help him adjust. My tasks were to connect with Charlie at a deep level to accustom him to living with humans; to read him accurately to enable him to feel more comfortable emotionally; to teach him to interact safely with humans and non-human animals; to help him adjust to life in a home; to socialize him, and to teach him that, even in captivity, life could be good. I also knew that it was vital to respect and honour his wild self. This was who he was, it was his intrinsic nature, and I knew that the personality which would ultimately emerge as he learned to relax would, most likely, always be underpinned by that essential wildness.

In his fascinating book *Animal Passions and Beastly Virtues: Reflections on Redecorating Nature*, scientist Marc Bekoff explains the difference between 'taming' and 'domesticating.' Dr Bekoff describes domestication as a slow evolutionary process that involves genetic changes, and which takes place under a degree of human control. This means that dogs, who have co-existed with us for over 15,000 years (and, depending on whether we include results

---

How was it for him? In the beginning it was not good

---

from mitochondrial DNA, possibly tens of thousands of years), are naturally classified as domesticated. Even feral dogs such as Charlie, who grew up without experiencing human interaction, possess genes which were changed through the previous domestication of their ancestors. Although these genes may evolve further if the feral dog comes from a long line of unsocialized feral dogs, they will never be the same as those of their wild forebears.

Yet all dogs – even domestic puppies – need to be socialized and taught how to live harmoniously alongside us. Leave any puppy born in a home to find his own way, without guidelines or boundaries, and havoc will ensue. Wildness will out. Wolves, the distant ancestors of dogs, can be 'tamed' in the sense of being socialized, but cannot be classified as domesticated, however well-socialized with humans they become. Their genetic makeup is not affected when they live in captivity because this evolutionary process takes many generations.

Suzanne Clothier, author, behaviourist, and founder of Relationship Centered Training™, reminds us to ask ourselves "How is it for you?" while working with, and even just being around, dogs. It applies just as well to any species we spend time with ... including our own. This question was one that I found myself asking constantly; and during the first few months, especially, the answers that Charlie gave me through his body language tore my heart to shreds, over and over again. How was it for him? In the beginning it was not good.

The slow process of helping our feral boy

... sanctioning the killing of all unowned dogs

adjust was hard for him for quite some time, with many setbacks along the way, and there were times when I questioned whether it was fair on Charlie to put him through this. It seemed a cruel thing to do: remove a feral creature from his natural environment, and expect him to learn to live happily in captivity. He had been taken from a difficult, perilous life, fraught with anxiety, danger and the constant threat of starvation; his scars told stories of fights with other canids, and injuries incurred while capturing prey. But it was the only life he had known, and he had his own social group who were familiar to him – especially Lenny, a small dog who was his constant companion, and who was captured alongside Charlie.

Most importantly, he had had the freedom to roam at will. In his new life he became a prisoner, unable to make choices and follow through on these; deprived of the ability to let his nose be his guide to new adventures, and run when his body was flooded with adrenaline and cortisol, the rank chemicals of fear. For a free soul who was most likely born of feral parents in the wild, to be captive must have been truly terrible, and he showed his feelings about this many times over during that first year.

Of course, Charlie was very much loved, well-fed, comfortable, and had excellent veterinary care, and I had a great deal of experience with traumatised dogs, fortunately, but for months it was clear that, given the choice, he would have opted for his old way of life, and would surely bolt for freedom, given the slightest opportunity. But if Charlie had remained in Romania, it's highly unlikely he would have survived more than a few more months at most. He arrived with a staphylococcus infection, demodectic mange and worms, and needed six weeks of intensive medication. Underweight, sick, and with a badly injured eye, his chances of surviving were very low. Also, Romania's policy to dogs changed that year, resulting in the countless cruel deaths of thousands of stray, street and feral dogs in that country.

Charlie arrived at my home at 7.30 in the evening of February 23, 2013. Seven months later, on September 25, the Romanian government passed a law sanctioning the killing of all unowned dogs after 14 days in the shelters. This came about after a child was attacked and, tragically, killed by street dogs in a Bucharest city park on September 2. No dog was safe in Romania until the law was suspended in June 2014, and the bodies of those who were brutally tortured and killed, then dumped, created mountains of shame in the streets. Even family dogs were torn from the arms of their carers, and rescue transporters who had already brought a number of dogs to the UK have subsequently imported thousands more of these at-risk animals to be homed. Charlie is very fortunate to be alive.

First Impressions

Gina Holmes, a friend who does a great deal of rescue work, collected Charlie from the transport van, and drove him to my home in a small village near Bath. The only information the rescue organization who imported him could pass on was that he was said to be 'a little nervous,' and they sent a rather fuzzy photograph taken while he lay on an operating table, waiting to have his left eye removed. In this, Charlie was staring directly at the photographer, ears pinned back, his strained face and body a visible map of tension, brows furrowed in a head that looked much too large for the rest of his body. His left eye was a mess, even after being cleaned up, and the pupil of his good, right eye was hugely dilated, rimmed by a pale amber-yellow iris. Although his shaggy yellowish, brown and cream coat looked similar to that of a German Shepherd, his head and large ruff appeared out of sync with his body, and his tightly pinned ears were floppy rather than pricked. He looked terrified and ready to bolt, and my heart went out to him.

Of course, at that time, I had no inkling of his background, and although I've fostered and adopted several dogs who were classed as cruelty cases, it was deeply upsetting to see just how traumatised Charlie was when Gina carried him indoors. He was smaller than I'd expected – the size of a small Labrador Retriever – the yellow coat in the photos was actually an unusual reddish-brown, auburn hue. His good eye was encircled by Cleopatra-style eyeliner, and a dark line transected the dome of his head, and spread down his back and tail, with an unusual, distinctive, saddle-shaped outline painted in darker fur over his shoulders and upper back (that I later discovered is a marking associated with wolves and wolfdogs). A thick dorsal cape (or ruff, as we generally term it) came to a clearly defined conclusion, creating a line of shorter fur beneath it, and his hips were narrow in comparison with his broad shoulders.

Gina had to rush home to care for her own dogs, so she set him down in the hallway, where Charlie stood frozen in place, paralysed with fear, for several hours. As the months passed, I was to see this freeze response occur time and again.

Skye, a beautiful, long-legged Deerhound/Greyhound/Saluki mix, has been a wise mentor and friend to many troubled and terminally ill dogs who have come to live with us during the eight years since I adopted him as a puppy. His trust and confidence in me has helped every dog to relax, and he has always somehow known exactly what each new dog needed, whether this be plenty of space to unwind, a playmate, or a friend to rest beside after surgery or during the final twilight days of life.

Skye was instrumental in helping Charlie to learn about life in a home, and that first evening he tiptoed across to Charlie, sniffed him cautiously, making sure to avoid coming too close, and stepped away, looking at me as if to say "This boy needs to rest." As Charlie had endured a two-day journey in a van, was unlikely to have been able to stomach food during that time, and was in a state of shock, I wholeheartedly agreed with Skye.

Helping Charlie to feel safe

After moving slowly to place a soft dog bed on the floor beside Charlie so that he could step onto it if he wished to, I sat on the floor in the living room, by the door to the hallway, taking care not to sit too close and invade his space. He could catch my scent from there, and I made sure to use peripheral vision for observing him. Direct eye contact is something that unfamiliar dogs avoid in engaging with each other as it is associated with issuing a challenge, and my priority was to help Charlie feel safe, while getting him accustomed to my presence.

Charlie stood perfectly still for hours, his

head turned towards the wall, bushy tail tightly tucked between his legs, blocking us out while signalling that he wasn't interested in interaction or conflict. I sat quietly, face and body turned away from him, occasionally speaking very softly

---

"Good boy" I whispered "Have some more"

---

to tell him he was safe with us. Skye went for a nap in the living room. Eventually, knowing how exhausted he must be, and hoping he was growing more familiar with my scent, I stood and moved slowly into the kitchen to gather a bowl of water and some chicken and rice that I'd prepared for his arrival. Sitting back in the same place, I turned my body sideways to show I meant no threat, and placed the bowls near Charlie. He shrank back, terrified, so I scooped some food into my hand and held it out near him. Charlie's head turned towards me, he stared directly at me with an intensity so powerful that it seemed he was looking deep into my soul, and after a few moments gently took a morsel of chicken.

"Good boy" I whispered. "Have some more." Charlie took another piece, and then carefully licked my hand as he dipped his muzzle down and ate. Using slow, fluid movements, I gathered up more food and he ate that, too. I raised the water bowl for him and he shrank back again. It struck me that perhaps he was unused to eating and drinking from bowls (which turned out to be true), so I dipped my hand into the bowl and gathered some water in my palm, offering it while it dripped through my fingers onto the rug. Charlie lapped it up carefully.

This was repeated with the food and water until Charlie's legs began to shake and buckle beneath him. He shot a sideways glance at me and slowly sank to the floor, face to the wall. I sat with him for a while, speaking quietly every now and again; then, careful to not make any sudden movements that could startle him, I rose and removed the bowls, and went into the living room so that Charlie could rest.

After a while his good eye closed. It was past one o'clock in the morning, and I motioned Skye to come upstairs with me so that we could all catch some sleep. Charlie opened his eye and shrank against the wall as we moved past him, but made no attempt to get up. When I came down to check on him at 5am, he was in the same position as when I went to bed. I expected to find a wet or soiled patch, but the rug was unblemished. Charlie had spent his first night in a home.

THree SeTTLING IN

Our garden angles down towards a narrow lane and a field on the edge of a stream, backed by woods. We walk the dogs in the woods, and a larger field a few hundred paces away, as a pony called Rabbit, and her young horse companion, Black Velvet, live in the field behind our house. Often, the two of them stand by the gate, waiting to say 'hello,' and hoping for a carrot or two. It took Charlie a long time to accept their presence, but now he simply glances across casually as we pass by.

This time, though, he's taken by surprise. Rabbit is grazing at the bottom of the steep slope, but Black Velvet is standing so quietly beside the gate that he only notices her while sniffing his way along the fence line. Charlie suddenly realises just how close Black Velvet is, and jumps, hops, and skips backwards, looking at me to gauge my reaction. I smile. "She caught you out today, my friend," I tell him. His eyebrows twitch in agreement, and he calmly walks beside me for the final few steps home.

On Charlie's first morning with us I left him to rest while I opened the back door for Skye, put food down for him, and made a cup of tea. Charlie's head didn't turn, but I could see his ears swivel as he listened to the unfamiliar sounds, although his face stayed turned towards the wall. Skye did his morning rounds, checking for new scents outside, ate his breakfast, and went through to the living room, stopping briefly to sniff at Charlie, who remained as still as a statue.

I brought through some chicken and rice and a bowl of water, and took up the same position as before – not close enough to be intimidating, but near enough to sit sideways and offer food in the palm of my hand. Charlie ate a little, glancing sideways, shrinking backwards after each morsel. He drank some water from my hand, and then, to my surprise, slowly, shakily rose to his feet and stepped into the dog

My heart ached for this poor soul who was in a new world

bed. Being in the middle of the hallway meant that we would have to step past him frequently throughout the day, which would make him even more anxious, so I lifted the bed, with Charlie in it, and carried him into the living room. Charlie didn't move when I set him down, and remained there until midday, when I took Skye for a walk. Charlie had not moved an inch when we arrived back home.

Charlie hadn't toileted at all by lunchtime, and he clearly didn't have the courage to leave his bed, so I carried him outside, bed and all, and set him down on the lawn, thinking that grass would be more familiar to him than the stone patio. After a few minutes he stood up and stepped onto the lawn, looking utterly terrified. My heart ached for this poor soul who was in a new world where nothing made sense. Charlie sniffed the grass and eliminated, then stood watching Skye trotting around the garden. After a while, as he hadn't moved, I took his bed inside, then gently carried him back into the living room and settled him.

He stayed in the same position all day, watching me constantly; suspiciously. I spoke quietly to him occasionally, and let him rest.

CAUTIOUS EXPLORATION

That evening, Charlie stood up and cautiously sniffed around the area close beside his bed. With body low to the floor, and tail tucked tightly between his legs, he slunk back and forth on his belly, investigating, flattening himself to the ground each time I shifted position slightly or glanced in his direction. I was careful to give him space and to avoid looking directly at him so that he wouldn't feel challenged, and I kept my movements slow and fluid. He ate a good-sized meal from my hand, and drank some water.

Again, I carried him outside to toilet, surprised that he could hold on for so long. It amazed me that Charlie toilet-trained himself instantly, clearly observing the rule of the wild that the den should be kept clean, and obviously viewing the house as his new den.

Charlie spent a little longer in the garden, though was too scared to come back inside on his own, so I gently carried him back to his bed. He sniffed the grass, stepping cautiously, watched Skye playing, and gave a small flick of the tail each time Skye approached him. They touched noses and Charlie licked Skye's mouth in a gesture of conciliation and respect. As the only company he had known in his life was that of other dogs, Skye's calm, gentle guidance proved indispensable in helping to build Charlie's confidence during those crucial early weeks.

FACING THE FIRST MONSTER

Two days after Charlie arrived I switched on the television, with the volume turned down to its lowest setting. A dog's hearing is far superior to ours, so the soundtrack which was almost

---

This was a monster that may attack him at any moment

---

inaudible to me was an assault on Charlie's senses. He had never heard music before, or seen digital images moving on a screen – he had no idea what a television was. Charlie's response was immediate. His head whipped up, his body became rigid, and he bolted out of the room.

I waited. A minute later he crawled back into the room on his belly, ears tucked back tightly, hackles up, and his eye rounded with fear.

Still crawling, he pulled himself along on his elbows to the dog bed with a face-on view of the screen and, stiff-bodied with tension, positioned himself so that he was up on his elbows. He eyeballed the screen furiously, and growled a deep, resonant challenge.

"It's okay, Charlie" I told him softly. "It won't hurt you."

Charlie didn't believe me. Why should he? To him, this was a monster that may attack him at any moment, and he was primed to retaliate. I kept very still, speaking quietly to him, and waited for him to figure out that this was nothing to be afraid of. He left the room several times, scuttering back each time with his tail tightly tucked between his legs, growling warningly at the screen. Eventually, he slunk to the other dog bed, furthest from the television, and looked at me with a confused expression, brows furrowed in an inverted 'V.' Slowly, I lowered myself to the floor beside him and, to my surprise, he leaned into me. We both gradually slid further down until I was lying on the floor with Charlie's back curled tight into my stomach. As I gently stroked his flank he relaxed and, after a while, fell asleep resting against me. This felt like an extraordinary privilege.

WHY THE FAMILIAR FEELS SAFEST

New experiences are terrifying for a feral dog, because he has no point of reference for them. The unfamiliar sets off danger signals, because anything at all could be a predator. In Charlie's mind, the television was potentially life-threatening. Nowadays, it doesn't bother him at all, and he shows a particular interest in watching CSI shows!

Not all feral dogs find domestic life so confusing and scary. Those who were born into domestic surroundings and then were abandoned, or who strayed and became lost, find it easier to readjust to life in a home. These are the dogs most commonly thought of when the term 'feral' is used: creatures who were once domesticated, but who have (for whatever reason) reverted to a wild state, and learned to support themselves independently of man. This type of feral dog will scavenge in areas

---

Feral dogs rely on fear to keep them safe

---

where there is human refuse. They're less wary of contact with people, but tend to keep their distance, if possible. I knew many dogs like this when I lived in Singapore and Malaysia as a teenager, and my sister and I befriended one particular dog who would approach us to say 'hello' each time she sighted us.

Charlie belongs to a very different type of feral dog, and to understand his nature and the behaviours I describe throughout this book involves some knowledge of why feral dogs behave differently to domestic dogs.

Charlie's kind are born to feral parents and raised in the wild. Self-sufficient, fearful of humans, and anything unfamiliar, they live by their wits. They battle to survive through hunting, scavenging, and keeping their instincts finely honed. Their mortality rate is high, their life span is shorter than that of domestic dogs, and their primary response to the unknown is fear. They live in close-knit social groups comprised of the mother, and possibly the father, as well as other dogs. The puppies forge neural connections in their brains for survival in their environment, but don't have the opportunities to form neural pathways that will help them to accept other species as part of their group – even other feral dogs outside the group may be treated with distrust or antagonism.

Whereas domestic dogs are promiscuous when in heat, and will mate with any available male, born-in-the-wild feral bitches select their mates according to how familiar they are, and may fiercely reject other interested males if they don't know them. In their paper, *Feralization: The Making of Wild Domestic Animals*, Thomas J Daniels and Marc Bekoff write of the definitions of feralization, including "the loss of tameness or approachability" and "the development of a fear response that essentially precludes subsequent positive interactions with humans." Essentially, the behaviour of these dogs is more like that of wolves than the dogs we are familiar with in everyday life.

Feral dogs are extremely fearful creatures – with good reason, because life for them is very harsh, and a powerful fear response is crucial to their survival. This intense emotional state is governed by the amygdala, an almond-shaped structure in the limbic system in the middle of the brain. The amygdala interprets information that is sent to the brain and translates it into emotions; specifically, fear and anger. It passes on this interpretation to other areas of the cortex so that hormones which set off an immediate response of freeze, flight or fight are released. You can thank your amygdala for the rush of adrenaline that gives you the sudden speed and strength to remove yourself from sticky situations. Charlie has demonstrated this to me countless times since he came to live with us, and the reaction each time occurs faster than a blink of his single eye.

Three factors are responsible for this heightened fear in feral dogs. The first – a very active amygdala – is chemical in nature, and aids survival of the fittest; the second, which predisposes a hard-working amygdala, is genetic, and the third factor is brain development based on early experience.

Genes affect behaviour because we

---

*Environments are enriched
through the company of people
and other animals*

---

all inherit certain characteristics from our predecessors. Fearful, extremely cautious parents are more likely to have fearful offspring – partly because of inherited characteristics, and partly because many attitudes and responses are learned from our parents. If you visit a litter of pups with a view to taking one home, it's wise to be cautious if the mother of the pups is nervous or aggressive, because you can guarantee that the puppies from that litter will grow up to be the same, however tenderly you love and care for them. The pups will have absorbed their mother's fearfulness both genetically and through observing, and learning from, her

responses. Dogs from puppy mills, born into a miserable environment bereft of stimulation and any form of socialization, are a clear example of the effect that deprivation has upon the amygdala and the psyche, as severe behaviour issues are common.

Feral dogs rely on fear to keep them safe. The amygdala works overtime, partly because this has been programmed into their DNA, and partly because they learn from their parents and social group that they must have swift reactions if they are to survive.

The third factor that makes feral dogs more fearful of new experiences and stimuli is their brain development. All our experiences from birth onwards, and our responses to these, determine how our brains will develop and function for the rest of our lives. In dogs, the first fourteen weeks define who they will become, and how they will react in maturity. This is the critical 'sensitive' period, when attitudes are formed for life. Every experience and stimulus creates neural pathways in the brain, forming new connections that need to be used to be maintained. You can imagine this as a path which you create in your garden. It has to be used and kept clear, or nature will take over and it will soon become overgrown and impossible to traverse. But unlike an overgrown path, if connections in the brain don't form, or become closed through lack of use, you can't cut back the brambles to clear them – you have to find a way to build another path.

If there are no opportunities for certain experiences, the neural pathways that enable negotiation of these aren't formed, and those potential abilities are lost forever. Puppies whose environments are enriched through the company of people and other animals, through frequent gentle handling, different sights, sounds and smells, sensory experiences such as grass, mud, water, carpeted floors, toys to play with, and games that engage their minds are fortunate, as they grow up well-equipped to cope confidently with new experiences. Because more connections have been made in the brain, they develop a more tolerant, sociable, confident, curious outlook on life.

Feral dogs born in the wild develop neural pathways that enable them to sniff out food and water, to form liaisons with their group members, to recognise potential threats to their safety, to hunt and forage, and to respond instantly to signals from the amygdala. However, as they live apart from humans, they aren't equipped to understand how to interact appropriately with people, or to cope with the mechanics of living a domestic life. Because of their evolution alongside humans, and the close attention they pay to our facial expressions and body language, domestic dogs read us well, and are able to follow our signals from an early age. Even puppies will follow the trajectory of a pointing finger to look at an object or be directed towards food.

With Charlie, this ability did not develop, presumably because there was no need for it in the wild. Later, once he had bonded closely with us, he learned to read us and follow cues, but this took several months of observing Skye's responses, and figuring out from those what was being asked of him. Some things, such as following the trajectory of a pointing finger, never have made sense to him.

Exploring Charlie's Heritage – or the other way around?

When Charlie had been with us for over eighteen months I decided to have his DNA tested. I was aware that, as he comes from Romania, some breeds from Eastern Europe wouldn't

be in the database, but I thought this would still be interesting to look into. Swabs taken from Charlie's mouth were sent to the Wisdom laboratory in the UK for breed tests. As Charlie's behaviour is so wolf-like, swabs were also sent to the Veterinary Genetics Laboratory at the University of California, Davis, to find out whether any wolf DNA had been present within the past three generations.

The results were extremely thought-provoking.

The Wisdom breed test showed no results for one side of his family tree. The other side flagged up a mixture of Japanese Chin, Standard Poodle, Pharoah Hound, Petit Basset Griffon Vendeen, Basenji, Smooth Fox Terrier, and Whippet. The wolf hybrid test indicated no wolf genes within the past three generations, though the laboratory noted that this didn't rule out wolf genes from further back in Charlie's heritage.

Charlie offers a glimpse into the distant past of some of our contemporary dog breeds

It's highly unlikely that Charlie is an actual mix of these breeds; rather that, as a feral dog, his DNA contains the genetic sequences that eventually were selected to create Japanese Chins, Poodles, and so on. Inheritance flows downstream, and so these breeds didn't necessarily contribute to Charlie's 'Charlie-ness' (as a friend put it on Facebook during a fascinating discussion about his DNA results). Instead, Charlie's heritage could be viewed as containing the seeds for what would eventually become other dog breeds: 'Japanese Chin-ness' and 'Poodle-ness.' Charlie offers a glimpse into the distant past of some of our contemporary dog breeds. Essentially, Charlie is, quite simply, Charlie!

Small Steps and Giant Leaps

On the third day Charlie followed me everywhere, keeping a safe distance, and instantly flattening to the ground if I turned to check out where he was. He began to drink from the water bowl I put out in the garden, and maintained this habit – although I have bowls of water in the living room and utility room, Charlie has never gone near them. However, he approached the metal bowl of food I set down in the living room that evening, then jumped back, terrified, when his nose touched it. I substituted it for an eco-bowl, which doesn't have the cold feel or the harsh sound of metal. Charlie approached it cautiously and I moved away. Immediately, he backed away, too, but looked longingly at the food. I sat a short distance from the bowl, placed some chicken in the palm of my hand, and offered it to him. He crept closer and took it, then ate the rest from the bowl. This felt like a huge breakthrough!

By the fourth day, Charlie was trailing me in a way that would have been comical, if not for his obvious anxiety and fear. He would crawl and squirm along, his belly touching the ground, a few steps behind me, stopping when I stopped, and moving each time I moved. If I went into the utility room to fetch something from the refrigerator, Charlie's face could be seen peeking anxiously around the doorway, brow furrowed, checking out where I was and what I was doing.

By this time he felt secure enough to go into the garden by himself, though he needed to be carried back indoors as he was too afraid of passing back through the doorway alone. Charlie followed Skye around like a puppy, returning Skye's play bows but holding the position of rear end up, shoulders and forelegs on the ground,

for several minutes at a time. He clearly had no idea how to proceed when Skye invited him to join in a game. Indoors, he lay close to Skye, not touching but clearly gaining comfort from his new companion's presence. Several times that evening Charlie approached me on his belly, using his front legs as levers to drag himself towards me, his toes and claws curling down to grasp the rug for additional traction. Each time he crept close I stayed very still, and spoke quietly to him. He stretched out to touch my hand with his nose, and allowed me to very gently stroke the underside of his face and neck. This sweet, frightened boy was reaching out, and I felt deeply moved and privileged.

During Charlie's first week with us he demonstrated what a quick learner he is. The challenges he had to overcome are described through the following chapters, but during those first few days I felt that, with patience, kindness and understanding, I would be able to help Charlie to overcome his fears. He and Skye began to play together on the seventh day, racing at full speed around the garden and giving huffs of delight, though I could only bring him back indoors by going out to catch him and carrying him inside. It made me feel guilty to have to do this, because he was clearly terrified of making the transition between outside and in, and the process of moving indoors threw him into a state of panic. This fear of doorways and narrow spaces was to continue for months.

Charlie began responding to his name by looking directly at me when I spoke it, and he started to leave his bed and greet me by dragging himself towards me on his belly when I came downstairs in the mornings. From observing Skye, he also learned to sit for a treat with no prompting on my part, and became increasingly more affectionate towards us: crawling over to

Charlie

put his head in my or Amber's hand, inviting a stroke or ear rub, and starting the day with a game in the living room with Skye.

Moment by moment, we were falling in love.

VISIT HUBBLE AND HATTIE ON THE WEB: WWW.HUBBLEANDHATTIE.COM
WWW.HUBBLEANDHATTIE.BLOGSPOT.CO.UK
• DETAILS OF ALL BOOKS • SPECIAL OFFERS • NEWSLETTER • NEW BOOK NEWS

Four TRUST

Charlie and Skye trot into the field, ears swivelling and noses twitching, checking out who has passed this way and left messages for other dogs to collect. They add their own olfactory signatures to the mix. Charlie's tail is high – a recent development – and he looks around confidently as he trips along beside me.

Suddenly, he stops and stiffens. His tail sinks down, and his eye grows larger and rounder as he glares at something on the grass some distance away. Someone, most likely one of the children who were picnicking in the woods yesterday, has left behind a white plastic bag, and Charlie has no frame of reference for this. He is frozen to the spot, gazing trancelike at the unfamiliar object.

I speak his name softly, but Charlie has slipped through the old portal of fear that opens up when he is faced with the unfamiliar. He doesn't hear me. A touch would only startle him and make him more afraid, so I call Skye to move in closer to us.

Skye's approach breaks the spell, and Charlie glances at him – then up at me – relaxing when I smile and tell him there's nothing to fear. We skirt around the bag, far enough away for Charlie to see it clearly without having to encounter it, and he takes a good look, glancing back at me to check that I'm not worried, sliding his head beneath my hand for a gentle, reassuring stroke behind the ears as we walk into the woods.

Observing Charlie during that first week prompted me to think a great deal about trust – the foundation for all healthy, meaningful relationships. As children we trust in the adults around us to keep us safe, to nurture us, and

Trust is a fragile flower: a gift that must be earned, appreciated, and never taken for granted

to meet our essential needs. If this trust is betrayed it leaves shards of pain: emotional scars that, though invisible to others, can eat away at future relationships. Our dogs, too, need to learn to trust us, and during this process we reciprocate by putting our trust in them. Shelters are overflowing with sad-eyed dogs whose trust in their carers has been broken, yet this can be rebuilt, given time, patience and love, and these are the dogs I find most rewarding to work with.

Trust is a fragile flower; a gift that must be earned, appreciated, and never taken for granted. It can be lost in a thoughtless moment; once lost, a great deal of work is necessary to rekindle it. Charlie had never had the opportunity to learn to trust humans. They were simply not part of his life and, if seen in the distance, were avoided, until humans began leaving food during a time when it must have been very scarce indeed. The Romanian winters are harsh, especially for a free-ranging dog. Many freeze and starve to death, as foraging and hunting bring little in the way of sustenance. To Charlie and his companions, these food gifts must have been very gratefully received. Ultimately, the price he paid for taking these was his freedom, and the prize gained was a longer, more comfortable life.

Rescuer or Jailer?

I was delighted that Charlie was willing to trust me, though this would take many months to fully develop. From his perspective I was his jailer in those early weeks and months, and he needed to gain my affection and guardianship in order to remain free from additional harm: a form of Stockholm syndrome, in which the captive bonds with the captor as a means of survival. The tender shoots that were put out during that first week began to take root, though there were many times when he reminded me just how fragile these were.

A change in my tone of voice, or a sudden movement instantly generated a fear response in Charlie, who would either flatten himself to the ground or bolt away and hide. Although he would eventually return, each time this happened it was at least two days before he was able to properly relax around me again, and we'd slip back to the sad sight of Charlie flattening to the floor at a random glance, or haring outside at the slightest movement. I had to carefully monitor myself at all times so as not to startle or worry him. My voice had to be soft – almost a whisper – and movements needed to be slow and fluid; not just around Charlie, but with my daughter and Skye, too. My relationship with Amber has always been good, but the extra care I needed to take because of Charlie added an extra dimension to this.

One of my previous foster dogs, Lulu, was a very sweet, black Greyhound who had suffered abuse before coming into rescue. The act of raising an arm to get dishes from the kitchen cupboard would send her into a state of panic, and if I extended a leg to step over an obstacle left by one of my children, Lulu would race into another room to hide. Eventually, she overcame her fear of being hit or kicked, and was happily rehomed, but I thought of her while working to teach Charlie that he was safe with me. The principle was the same: earn his trust by proving to him that I was worthy of it.

I made mistakes, of course. Moving too quickly to answer the door when the postman

---

I didn't want any of those fates for Charlie

---

knocked, running to close the back gate which had not been properly bolted when it flew open in a gale, and running down the stairs when the phone rang were instinctive responses for me. Charlie's reaction each time was just as instinctive, however, with much low-bodied running back and forth, and we had numerous setbacks. I had to choose to miss the postman and return phone calls later, though the gate needed immediate action simply to keep Charlie from disappearing over the horizon. Sadly, a lot of imported street dogs have escaped after arrival in the UK, through slipping their collars, jumping out of cars, or bolting through a door left ajar. Some were never seen again, others died on the roads, and the lucky ones were tracked down and recaptured. I didn't want any of those fates for Charlie.

The phone posed a huge challenge for Charlie for several months. Even though I lowered the ring tone to a pitch that I could only just hear, Charlie would howl the moment he heard it, and only quieten after what he clearly considered to be a horrible noise ceased. His reaction turned out to be quite useful as an alert when I was in another room and didn't hear the phone ring!

Meeting new people

Visitors were kept to a minimum for quite a while. My veterinary surgeon and close friend, Amelia Welham, came to the house to check Charlie over a few days after he arrived. Amelia has cared for all my dogs and foster dogs over the past few years, and treated Skye and four of my elderly, terminally ill fostered and adopted dogs during that time. Her compassion and patience are extraordinary, and I knew that she would be extra-careful while handling Charlie – who flattened to the floor as soon as she arrived, and lay tense but still while she examined him. I knew he had worms from cleaning up after him in the garden, and I was concerned about some bald patches and very sore skin on his throat and the tops of his forelegs.

There's a huge difference in attitude between dog lovers and dog people

---

Amelia was sure he had demodectic mange, which is caused by demodex mites, and a staphylococcus skin infection, and she sedated him briefly in order to take skin scrapes. Scrapes are very painful to endure as they involve removing the top layers of skin, and it would have been cruel, given Charlie's level of trauma, to put him through the procedure without medical help. By the time the results came back confirming her suspicions, Charlie's entire neck was a mess, red raw and bleeding. Amelia put him on six weeks' medication, however, and he healed well.

Previously, Amber's friends, when visiting, had always popped their heads around the living room door to say hello to me, but after Charlie arrived they went straight up to her room, as Charlie bolted to the garden in fear each time they did. Often Skye, who's immensely sociable, would run upstairs to join the young people, revelling in the attention they gave him. My friends visited one or two at a time, and I spoke to each beforehand to prepare them for meeting Charlie. Our new house rules were: make the usual fuss of Skye (who has quite a fan club), avoid looking directly at Charlie, and on absolutely no account step towards him or try to stroke him. Any attempt to do this instantly sent him into panic mode, and I had to teach him to feel safe around people. Happily, only one visitor refused to follow The Charlie Rules, as we called them.

There's a huge difference in attitude between dog *lovers* and dog *people*. Dog lovers really, really like dogs, and want to make a fuss of every dog they encounter, regardless of whether or not that dog wants to make friends. This puts pressure on the dogs they meet if they insist on stroking them, even when the dog's body language is screaming "Please leave me alone – I'm scared!" and they're convinced that every dog will be grateful for the attention they're offering.

Dog people love dogs, too, but they also respect them. They take care to understand them, and interpret the signals the dogs are giving out. If a dog needs space, dog people give this, however cute and cuddlesome that dog looks. They wait for a dog to approach them and seek out contact, and they're respectful in how they speak to, and touch, dogs. Pats on the head are a no-no, as the dog loses track of where the hand is, and can therefore feel threatened.

Charlie

Gentle strokes under the chin or along the flanks are friendly gestures as far as a dog is concerned. Dog people know this.

The last thing Charlie needed around him was a dog lover, though dog people in small doses were welcome while I helped him learn to socialise. One friendship came to an end because I refused to allow him to be subjected to unwelcome attention: it simply became too stressful for both Charlie and I, despite me constantly explaining The Charlie Rules before each visit. And it tested my dance moves, having to continually step in-between them while she dove forward, arms extended, shrieking "He's so sweet – I have to cuddle him!" while Charlie flattened to the floor briefly before turning tail and fleeing to hide in the garden as far away as possible. Each time, there was the customary two-day recovery period for Charlie, and it simply wasn't fair on him.

However, other friends were very respectful of Charlie's needs, and were happy to allow him space to adjust to their presence. Something else that came in useful was small pieces of cheese and a jar of liver treats. Charlie and Skye like liver treats very much, but after a while we discovered that Charlie, as well as Skye, would do just about anything for cheese, which had almost magical powers where these two were concerned.

Each guest was handed a few treats and small chunks of cheese as they came through the door. Skye would be right there, waiting to say hello, whilst Charlie hung back behind him, preparing to flee if anyone so much as glanced at him. While Skye was being greeted, some of the goodies were casually but carefully thrown away from Charlie, so that he had to retreat towards the kitchen to eat them. This helped him feel safe, as well as getting him his favourite food rewards,

because he wasn't put in a position of having to either step forward or bolt. It also gave plenty of space for my friends to step into the living room, joyously accompanied by Skye, without the risk of walking directly past Charlie.

A dog's perception of the world is very different to our own

Initially, Charlie would creep back to peek through the doorway, give the canine equivalent of a gulp of dismay at the sight of additional people on the sofas, and then beat a hasty retreat to the garden. After a few weeks, however, he began to stand close behind Skye, gently wagging his tail as friends walked in. This graduated to coming into the living room using his customary low-bodied crawl and slink, and sniffing cautiously at our visitors – who politely looked away and made no attempt to touch him. The day Charlie first nudged his head under a friend's hand to invite a stroke we had a necessarily quiet, but nevertheless jubilant, celebration!

Nowadays, Charlie steps ahead of Skye so that he's the first to greet friends as they arrive, his tail wagging madly as he performs a happy dance. He actively seeks out attention, and will offer a paw over and over again, placing it on the lap of his chosen friend and curling his toes and claws to grasp a knee or hand. Charlie's paws are large in proportion to his body, and he uses them in a similar way to a cat: presumably, this was one of his strategies for catching and holding on to prey in the wild.

The nose knows

Some fourteen months after Charlie's arrival, though, his reaction was quite different. I'd attended a fascinating, three-day course in the area, titled An Holistic Approach to

Understanding, Communicating and Healing Your Dog: Creating Mutual Trust and Respect, taught by wolf and dog expert Dr Isla Fishburn. At the end of the third day, Isla, our friends Stef and Leon, and my vet friend Amelia went for a meal at The Globe, a lovely old pub near the village I live in. After a happy couple of hours excitedly discussing aspects of the course and sharing our own experiences of working with very troubled dogs, we came back to my house.

Charlie and Skye were delighted to see everyone, and Charlie, as always, burrowed his nose deep to sniff out who we'd all been consorting with, canine-wise, as we moved into the living room. Suddenly, he leaped back and ran into the kitchen, his body tense, then crept forward to peek through the door at us. After a while he came in, sitting close to me and offering a paw, then left and returned several times.

At first I thought perhaps it was too many people arriving at once – but as a group of seven of my canine behaviour students had recently visited for their practical study day with me, and Charlie had been delighted to have them around, his reaction was surprising. Only after everyone had gone did it occur to me that as Isla, Stef and Leon all live with wolfdogs, they smelled very different to my other friends with domestic dogs. Charlie most likely encountered wolves in his previous wild life, as there are comparatively large numbers of them in Romania and the Carpathian Mountains. Carpathian Shepherd Dogs are bred to protect livestock from wolves and bears, and Isla had commented that Charlie looked incredibly similar in size, colour and coat pattern to a Carpathian Shepherd/wolf mix dog she had worked with.

The most important aspect of my work with dogs is to figure out what is going on inside their heads, which is why I trained in canine psychology – an area that holds tremendous fascination for me. It can be hard enough for us to work out what other humans are thinking, and what their motivations are, and to attempt (and presume) to delve into the mind of a creature of a different species can open up intriguing new horizons.

A dog's perception of the world is very different to our own, and they experience the environment through their extraordinary senses in ways that we cannot possibly imagine. Yet, really, dogs are transparently easy to read and understand when we know how to do this. They make no attempt to hide their emotions – what you see is what you get – and their main form of communication is body language. By observing their responses to particular situations and stimuli, we can reach in and follow the clues that they offer so eloquently, enabling us to decipher reasons for specific behaviours, and look for ways in which to compassionately work with these.

Looking at the world through Charlie's golden-yellow eye has helped me understand him better and protect him. Because he now knows that he is safe with me, Charlie's trust has blossomed to the extent that I am a rock he can lean on in times of uncertainty and fear – just as he came to me for reassurance after picking up the confusing scent of wolfdogs from my friends.

FIVE   BONDING AND TEACHING

Just over a year after Charlie's arrival, more evidence of his powerful bond with Skye surprises me. Both dogs love cheese, and Charlie has a sixth sense that always alerts him when I take some from the fridge to grate for a meal. He's standing just inches away from me before the package is even opened, his amber eye bright with hope as he licks his lips in eager anticipation. I cut off two small slices, one each for him and Skye – but instead of stepping forward to get his piece, Charlie turns around and trots into the living room where Skye is sleeping.

A moment later he reappears with Skye close behind. They both sit side by side, waiting, and Charlie repeatedly casts joyful glances at Skye, his tail thumping on the ground behind him, as if to say "See, I told you she has cheese!"

This has happened every time since, and it's heart-warming to see how Charlie is so willing to share that he'll go to find his friend.

Skye is a natural mentor. His calm presence and patience, coupled with his willingness to play if invited, has reassured many traumatised dogs who have come to live with us over the past seven years. His bond with Charlie was immediate, and he's been my helpmate through even the most challenging times with our feral boy, and the onset of aggression issues – seven months after Charlie came to us – was a tough time for all of us.

Charlie's selfless determination to make sure that Skye shared in the cheese illustrates how much he loves his friend. To be honest, if Skye had been first in the kitchen, he would

___

Skye is a natural mentor

___

have happily gulped down all of the cheese without a thought for Charlie. Often, even lifelong companions will ensure that they are first in the queue for a tasty treat, and resource-guarding over food, toys or attention is a common issue among dogs who otherwise get on well.

Fortunately, there's only ever been one bout of (one-sided) conflict over a possession, and Charlie had been with us for ten months by then. On Christmas Day, 2012, I gave the boys each a large, extra-strong, bone-shaped Kong™. This is a hollow dog toy that can be filled with food to keep dogs mentally busy while they work to get to the goodies inside. There are several types of Kong, but the ones I bought needed to be heavy-duty because Charlie's jaws are so powerful he can bite straight through a tree branch in one strike, and no other dog toys have survived more than a minute in his possession.

I stuffed the boys' Kongs with liver treats and cheese, and handed them over. They happily carried off their gifts – Charlie to the living room and Skye to the garden – and got to work excavating the contents.

A while later Skye trotted through, Kong in his mouth, whereupon Charlie growled and hurled himself at Skye, who instantly dropped his Kong and retreated. Charlie grabbed the Kong, put it beside his, and stood over both prizes, hackles raised, fiercely growling. Clearly, he considered the Kongs to be so precious that no-one would be allowed near them – and as he was in the centre of the room, he was effectively guarding the entire room. Sneaky measures were needed to put an end to the hostile atmosphere that was permeating the house.

I went into the kitchen and got out a pack of cheese. Within seconds, Charlie was right there, sitting beautifully and looking expectant. Skye hovered in the doorway, anxious about Charlie's unexpected hostility, but Charlie had already forgotten the Kong incident, and glanced back as if to say "Hey, come on – she has cheese!"

Skye looked at him disdainfully, as if to say "What? Sit next to you when you just rushed me? No thanks!"

In the meantime, Amber slipped into the living room, collected the Kongs, and hid them away in a drawer. The boys followed me back through after their treat; Charlie sniffed around a little, puzzled that his treasures had vanished, and they both settled down on their beds. Peace reigned once more, and the Kongs were banished forever.

FORMING BONDS

Charlie's way of bonding with other dogs is through play, and he plays rough in comparison to domestic dogs. As you'll read in chapter 10, he initially displayed a powerful hatred and mistrust of all dogs except Skye, and I had to do a lot of work, teaching him to socialise safely. He also developed very bullying behaviour towards people for a while, particularly if he sensed any form of weakness, and remedying this involved

careful effort. However, Skye was Charlie's only familiar point of reference when he arrived at our home. He didn't understand humans, and was deeply afraid of them, but he took comfort from the presence of a dog who was relaxed, willing to give him space when he most needed it, and who is an exemplary guide and teacher.

Each morning began with the canine version of a high-five

Soon, Charlie became Skye's shadow. He followed him around the ground floor of the house and the garden, though couldn't pluck up the courage to enter the utility room or Amber's studio at the bottom of the garden, and he seemed totally unaware that the stairs even existed during the first six months. If Skye lay down to rest, so did Charlie; when Skye got up to go outside, Charlie was so close behind that he risked stepping on Skye's rear paws.

Then, one day, after several weeks of play bow invitations that went no further than a view of airborne rear ends and wafting tails, Charlie sprang up, ran around the garden at full speed, swiftly followed and overtaken by Skye, and hurtled directly towards Skye as he turned to negotiate another lap. Skye jumped sideways at the last moment and both dogs slid to a halt, their paws leaving long skid marks in the grass while clods of earth flew up into the air. (Fortunately, I've become far less attached to the idea of having a tidy garden since Charlie arrived, as one of his favourite outdoor occupations has turned out to be digging impressively deep dens in the middle of the lawn!)

Charlie leapt forward on his hind legs, forelegs pounding the air. Skye followed suit, and they engaged in a full-on boxing match, twisting and turning with tails wagging madly and mouths open in happy, doggie grins. The boxing alternated with sudden mad dashes and tag games, each switchover signalled by a momentary pause and excited bark. Skye's energy ran out first. Being a Lurcher he's a sprinter, built for speed, and until recently has outrun every dog who's been willing to engage in a race, but endurance isn't one of his strong points. Skye shook himself to signal the end of play, and went to take a drink of water from the outside bowl. Charlie followed suit, and they then lay down on the lawn, side by side, panting and looking pleased with themselves.

From that day on, each morning began with the canine version of a high-five, followed by a boxing or wrestling match. Additional moves were added for extra fun. Charlie would run underneath Skye's belly, the difference in their sizes making this only just possible. Skye would take a flying leap right over Charlie's back as Charlie twisted to see where his friend would land so that the next stage of the game could begin. It was far more entertaining to watch than any television show.

Charlie's bond with Skye served to boost his confidence. He took his cues from Skye's relaxed attitude, and even though some situations may always be too much for him to cope with, he learned to accept, and even to enjoy, many experiences that had initially terrified him. Grooming was one of these.

Introducing Grooming

Skye has a long, rough coat, thanks to his Deerhound heritage, which requires regular brushing to avoid the formation of dreadlocks on his haunches. Being groomed is one of life's pleasures for Skye, who'll happily stand still for as long as it takes, a blissful smile on his face, and he knows that a treat will be forthcoming once his coat's been attended to. A hose down after he's swum in the stream and emerged wearing a cloak of algae isn't as welcome, but he patiently endures it.

A children's paddling pool - the perfect dog bath!

Charlie had a shaggy coat and ruff when he arrived, though he gradually shed it all and emerged with much shorter fur. The shedding was extreme and incredibly messy. Charlie's passage through the house and garden could easily be tracked by the vast quantities of fur that floated along behind him, drifting everywhere like an excess of tumbleweed from a spaghetti western. One of the rugs in the living room seemed to have a Velcro™ effect, and Charlie only had to walk across it to leave a thick layer of auburn, brown and cream fur across its surface. Also, he smelled pretty rancid, and his coat was so greasy that we could actually see the oil on our hands after stroking him. Something had to be done.

Amelia called round with some dog shampoo for sensitive skin, as Charlie's staphylococcus and demodex infections had made his skin very sore. In preparation of a bath, I tried to carry Charlie upstairs, but, two steps up, realised that it was one thing to carry a 50 pound dog (he was, at that time, still underweight) in from the garden at the end of the day, but quite another to negotiate steep stairs with him in my arms – especially as Charlie's chosen form of resistance was to go limp and become a dead weight. Clearly, another strategy was needed.

Fortunately, it was a hot summer's day and Amber (then nineteen years old), deciding she'd rather paddle in the privacy of our garden than down in the stream, had bought herself a large children's paddling pool – the perfect dog bath! – with extra buckets of water carried out for rinsing. As soon as Charlie realised that I intended to coax him into the pool, he ran off, which meant I had to catch him – never an easy task – and he cringed and cowered throughout the bathtime ordeal. I felt really sorry for him, but the deed was done as quickly as possible, although I was as drenched as he was by the time I released him.

Leaping out of the pool, Charlie vigorously shook himself, raced around the garden at top speed with Skye, rolled all over the grass ... and refused to associate with me for the rest of that day. I didn't blame him.

I use the gentle Sympatico method that I devised years ago

The next stage was grooming, to remove as much of Charlie's dead hair as possible. There's a handy little gadget called a FURminator®, which can be over-excessively used when stripping dogs with very thick coats (I prefer a dog rake for this purpose), but which was necessary for Charlie. Skye posed beautifully as the model example of the grooming process, though I substituted his double-sided brush and kept the FURminator in my pocket so that I could use sleight of hand to make a crafty switch when it was Charlie's turn. Charlie stood at a safe distance and watched.

As soon as I finished, I offered Skye a piece of cheese from the small pile I'd secreted in my pocket with the FURminator. Charlie stepped forward, his golden-yellow eye glowing. I held out more cheese, and lightly touched Charlie's back with the FURminator as he took the cheese. He stood quietly, waiting for more. I repeated this several times, gently moving the gadget along his

back without actually using it, then walked away. A while later I repeated the process several more times, slipping the FURminator into place, and stripping away just a little fur each time. It took several days, but soon Charlie would come and stand close when he saw me prepare to groom Skye, and nowadays he leaps around, butting into Skye's grooming session, wanting to be first!

Learning good manners

Charlie had a lot to learn about living in a home. I've fostered and adopted several dogs who were lacking in house manners. Some were ex-racing Greyhounds who had lived all their lives in kennels, and some were dogs who hadn't been properly taught, and whose subsequently appalling manners had led to them being either abandoned or relinquished to a rescue.

The first lesson in polite behaviour nearly always concerns food: mine, as well as theirs. It's not good for the digestion to have to dodge and dance on the sofa for each morsel while a determined dog (who has always been fed before us to eliminate hunger pangs) mugs me for my dinner. I was grateful that Charlie, unlike the other dogs, never attempted kitchen counter-surfing, but he would come and place his paws on my knee, toes curled; claws digging in, while he loomed over me in an attempt to snaffle my food.

The solution was simple and effective. Each time Charlie made a beeline for me with a gleam in his eye and intent to help himself writ clearly all over his face, I turned sideways, shielding the plate with my arm, and picked up a book, opening it to create a barrier between him and the food. As I have a habit of reading during meals when alone, it suited me just fine, and also blocked out Charlie's unwelcome attentions beautifully. Within four days he was following

Charlie

Skye's good example and going to lie on his bed as soon as I brought my meal into the room.

Although, during the first couple of months, Charlie continued to hide away from visitors who were unknown to him, he began to come out of his shell with familiar people. Gina, who had brought Charlie to us on his arrival in England, came to visit a month later, and was astounded at the transformation. Cautiously, Charlie stood behind Skye when she came through the door, then suddenly became very alert when he recognised her scent, and ran over to greet her, his tail wagging. He stayed close to her, repeatedly placing a paw on her lap, until she left. It helped even more that Gina brought along a large bag full of home-made liver cake, and handed over some of it to both boys before I stored the rest in the fridge and freezer.

---

*He didn't see why he should
come inside*

At around that time Oliver, the second oldest of my four sons, came to spend the day with us before heading off to travel around the Far East. Charlie was very nervous of him, and hid away whilst Olly was there. Four months later, Olly returned to the UK and came to stay the night. Charlie remembered him, and decided that Olly was his new best friend, sitting close by and gazing at him adoringly.

First steps in recall

All dogs need a degree of training, which I view as simply teaching them the responses that make them a pleasure to be around, and which are essential for keeping them, humans, and other animals safe. Although I have trainer friends whose dogs can perform amazing feats, my dogs are simply taught the basics that correspond with my house rules, and I use the gentle Sympatico method that I devised years ago, which is taught at the International School for Canine Psychology & Behaviour, a college I founded. This method involves creating a mutual bond of trust and affection through being considerate of each dog's individual needs, and it has helped many abused and traumatised dogs to recover from their sad pasts.

Good recall is vital. So is waiting when asked; polite manners when meeting guests and other dogs, and table manners. Charlie soon learned to respond to his name, and I made this into a game of hide-and-seek between downstairs rooms as a way of teaching him to come when called. He thoroughly enjoyed this, especially as each time he came to me he was rewarded with a tasty treat and fulsome praise.

Persuading Charlie to come in from the garden necessitated a lengthy process over several months because of his extreme fear of doorways and corridors, and he had to negotiate both to come inside. Inevitably, I'd have to go out and call him to me when he went in the garden with Skye at night, then pick him up and carry him indoors. As his weight was steadily increasing, this was hard work – especially on nights when he led me a merry dance around the garden first, because he didn't see why he should come inside. There are no street lamps in our village, so it really is pitch black at night, and a dog hiding by the hedge or behind the lavender bushes is very hard to see.

About five months after he arrived, Charlie started to respond to being called at night, and would dash to the back door, wait while I retreated to the living room, then run through to join me. I'd then give him a treat and sidle carefully past him to close the door before he disappeared outside again. If anyone was foolish enough to stand in

continued page 41

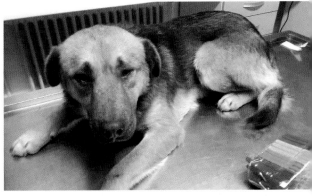

Above left: Charlie with his constant companion, Lenny, in the wild in 2011, before Charlie lost his left eye. Denisa, their rescuer, took this from a distance, zooming in to get a clear view of both dogs.

Above right: Charlie flattens to the ground in terror during his capture; a position he subsequently adopted with me whenever he was very afraid. The damage to his left eye is clearly visible.

Terrified and in pain. Charlie in Romania, about to undergo surgery to remove his damaged eye.

Charlie in Gina's car, huddling against another rescued dog for comfort.

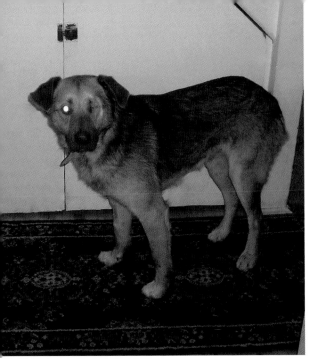

On arrival Charlie stood for literally hours in the same spot in the hallway ...

... then, finally, lay down with his face averted.

It took a while but eventually Charlie learned how nice it was to sleep on a bed!

*Charlie feels brave
enough to begin exploring
the living room ...*

*... though is still too nervous to
investigate the garden.*

Picture Gallery

Charlie wears his Perfect Fit harness for the first time.

Though very wary, Charlie takes his cue from Skye's relaxed attitude.

Learning to feel comfortable in the field by focusing on me and the treats.

First lesson in loose leash walking.

Skye watches over Charlie.

The first pavement walk.

Picture Gallery

Learning to sit in the field. Lots of treats helped!

The first exploratory sniff in the field.

Charlie begins to interact and play with Skye.

Mayzie comes to play with the boys.

Charlie during his most challenging time, seven
months after arrival.

Making friends with Cyder.

A proud moment. Charlie has crossed the bridge
over the stream for the first time.

The bridge that caused Charlie such fear and consternation.

Charlie learning to wait for the ball.

Playful Charlie.
(Courtesy Kerry James)

Making friends with Rosie and Caroline.

the kitchen or hallway, he would panic and run back outside, and the 'catch Charlie' game would have to be played because no way was he going to come back in. Now, he happily races indoors at top speed when I call him from anywhere in the house, and will blithely trip past me, even if I'm standing by the back door.

Broadening Horizons

The next step was teaching Charlie to accept and cope with wearing a harness and leash. As a free-ranging dog, he had had the freedom to go wherever he wished; whenever he wanted, and had never worn a collar. I put a collar on him the day after he arrived, so that he could wear an identification tag with my contact details in case he found a way to escape. However, walking a dog with the leash attached to a collar can cause numerous health problems if the dog pulls, so I ordered a Perfect Fit harness, which is what I recommend to my clients. The harness comprises three separate, adjustable parts for the back, chest and girth, and is very comfortable to wear. I fitted Charlie's harness, and left it on him so that he could become accustomed to the unfamiliar sensation. He cringed when I put it on, but soon relaxed and went to his bed to rest. Later, I removed the harness, and, over the next few days, put it back on him for short periods to get him used to it.

For Charlie's first experience of being on-leash I filled my pocket with treats, slung a leash around my neck, and took him and Skye out into the garden. Charlie followed me around, beautifully focused on my pocket, while I occasionally fed him treats. I clipped the leash to the back of the harness. Charlie cringed, but relaxed when I let go of the leash so that it could trail behind him as he followed me. After a few minutes I picked up the leash but kept it very

loose whilst walking around the garden. Charlie stayed close by, and every so often I signalled to both of them to wait for a moment by standing still. We repeated this exercise, briefly, several times a day for just over a week.

By the end of this period Charlie was obviously feeling secure and comfortable in the garden, so it was time to introduce him to a wider view of his new world. Amber took Skye's leash, I took Charlie's, and Amber opened the back gate so that Charlie and I could follow her and Skye. Charlie stepped through the portal, froze, and flattened to the ground. Immediately I retreated, calling him as I rustled my 'treat pocket,' and he rose and followed me back inside the garden. We practiced this twice daily, and Charlie took a step further beyond the boundary each time. It took another week to travel the thirty or so paces to the edge of the lane, but after that Charlie was eager to go out with us.

Going to the field, about a hundred yards away, was harder for him, and it took some time before he was able to walk that distance. Although we live in a quiet village, there are cars, vans, children on scooters and bicycles, and the occasional motorbike passing by. Just the sight of any of these in the distance prompted Charlie to hurl himself flat on the ground and freeze, so I chose the times when it was unlikely that much would be going on.

Finally, after numerous attempts, we made it to the field, and Charlie watched in amazement as Skye ran about, chasing his favourite ball. The slightest unfamiliar sight or sound caused Charlie to either freeze or attempt to bolt: a drifting leaf, a bird flying past, children's voices, all provoked the same reaction. It was months before Charlie could happily trot around the field and woods, and, even now, anything unusual still provokes a freeze response. Over

charlie

and over again I thanked my lucky stars that we didn't live in a city!

Charlie's manner of locomotion is very different to that of other dogs. Dogs bounce, to varying degrees, as they trot and run, because a dog's paws flex slightly with each step, creating a shift in the line of the shoulders and back. Charlie has large paws in proportion to his body size, and he moves like a wolf, with no flexure of these. His legs extend in a straight line, back and forth in a flowing movement that appears to carry him over the ground effortlessly. Seen from a distance, he seems almost to float, and his gracefulness is truly beautiful to witness. His natural gait is equivalent to my jogging pace, and he loves to run alongside me as I jog, though he will slow down if I need to walk. Try as I may, I would never be able to emulate the sheer magnificence of his smooth, easy stride. Outside, in his natural element, Charlie embodies true poetry in motion. His wild self, never far beneath the surface, emerges in all its glory.

VISIT HUBBLE and HATTIE ON THE WEB: WWW.HUBBLEandHATTIE.COM
WWW.HUBBLEandHATTIE.BLOGSPOT.CO.UK
• DETAILS OF ALL BOOKS • SPECIAL OFFERS • NEWSLETTER • NEW BOOK NEWS

SIX RESPONSIBILITY

It's a warm, sunny day, and Charlie is lying on the patio just outside the back door, engaged in his favourite pastime of being king of all he surveys. From here he gets a good view of the garden, and can also sense activity from inside the house. Nothing escapes his keen attention.

Charlie's head is raised and he rests regally on his elbows with his back curved, bushy tail curled loosely around his body. There's a feeling of calm and stillness about him as he inhales the scents that drift through the soft air, and listens to sounds that are beyond the range of my comparatively dull human ears. He radiates relaxed awareness, enjoying the moment, though the slightest unfamiliar sound will bring him to his feet, hackles up, as he prepares to defend his territory.

Taking care of another living creature, whether human or non-human, necessitates accepting responsibility for their wellbeing and safety. The very expression 'take care of' indicates the process of caring for, looking after, concerning ourselves with, nurturing, and helping that soul to achieve full potential. It's a term of selfless love; or rather, it should be. Whether our care is extended to children, family members, friends in need, those we are employed to help professionally, or non-human charges, it has to be unconditional. Putting conditions on caring taints and ultimately undermines the relationship.

Choosing to take responsibility involves accepting the likelihood that there will be a degree of sacrifice. Because it involves doing whatever is best for the other party, this can result in making decisions that cause us inconvenience, either minor or major. Charlie has taught me a great deal about the absolute need to accept and accommodate this.

He radiates relaxed awareness

Offering to foster Charlie was my choice, though it was an uninformed one at the time because I had no inkling of his past. Adopting him was a decision I made based on a two-week acquaintance, in the full knowledge that I would need to make some major lifestyle changes in order to help our feral boy learn to cope with domestic life. The adjustments I made to accommodate his needs were small fry in comparison to those that Charlie was undergoing, but they had a significant impact on our lifestyle for a long time. It's not an undertaking for the faint-hearted, and although an understanding of dog psychology and behaviour is important for all dog guardians, it really is vital when living with a free-ranging dog, or a dog with serious issues. This is not a decision that should be made lightly, and most certainly not because of a romantic notion of life alongside a previously wild creature, as other fosterers and adopters of free-ranging dogs have found to their cost – and that of the dogs. In some cases those dogs sent to inexperienced homes have paid the price with their lives ...

When I signed the adoption form and returned it to the rescue, I also signed a contract that was visible only to me, written indelibly on my heart. This private contract included three promises: to give Charlie the best possible life I could offer; to make every effort to help him learn to live happily in a world he found frightening and confusing; to honour and respect his nature.

Fortunately, I lead a fairly quiet life, and Amber was never interested in hosting wild teenage parties, so it was easy to establish the foundation for settling in Charlie and helping him begin to overcome his fear of people. This meant that I was careful about who visited us, because loud voices and voluminous bodily gestures threw Charlie into a state of panic. Because he was so transparently terrified of company, some friends were nervous of him, and afraid he may react aggressively. This never happened when he was afraid, but did occur later when he gained confidence, and a bullying streak emerged.

My social life narrowed: I could go out only rarely, and trips away were cancelled, but the friends who visited were very considerate of Charlie's needs. These included Amelia, of course, who was providing Charlie's veterinary care; my close friend and colleague Theo Stewart, aka The Dog Lady, who lives four hours' drive away, but visited several times, and we spoke on the phone at least twice weekly; Lisa Dickinson, one of my behaviour students who became a close friend; Carole Negre and Yolanda Cheung, friends of some ten years' standing; my cousin, Sue Beech, and Shirley, a friend from the village. Charlie soon grew to enjoy their company, and when he began to greet them at the door with Skye, and choose to come and sit by them, they each viewed this as a gift and a compliment.

Stimulation had to be kept to a minimum, and new people and experiences were introduced very slowly and carefully, with attention to Charlie's body language to make sure he felt comfortable. Charlie got used to the television, and the volume was very gradually increased to correspond with his ability to stay in the room if it was switched on at night, but it took a long time for him to stop reacting to music, even at low volume. His initial impulse was to turn tail and hide, which graduated to his remaining in the room but howling at full volume, as he also did when the phone rang, when people walked past the garden, and when he heard a dog barking in the distance. Dancing was banished for months, though after a year he not only coped with this but also decided to join in!

---

*... so harmony in the home
reached a new level*

I gave up playing my guitar for almost a year – something I'd regularly engaged in before Charlie arrived – because his reaction was first to howl, and then attempt to attack the guitar, hackles up and teeth exposed. Clearly, he didn't appreciate my musical skills! In the interests of a calm atmosphere and an intact guitar and fingers, I took it from its case once a month, finger-picked softly for a couple of minutes, crooned sotto voce if Charlie stayed on the other side of the room, then put the guitar away again. Over the course of the year my voice grew rusty and my fingers stiff, but Charlie eventually decided that, actually, being serenaded was quite a pleasant experience. He would come to sit beside me, looking for all the world as if he was enjoying himself. This new liking for music fortunately expanded to include CDs, so harmony in the home reached a new level.

WHO'S PROTECTING WHO?

We live at the far end of a small road in our village. The front garden was bordered by a low hedge and metal gate when Charlie arrived, and the back garden is mostly surrounded by a very tall hedge, with a fence bordering our neighbours' garden. Each time my neighbours parked their van in front of our house and walked across in view of the living room, Charlie was convinced that his life (and possibly mine, too,) was in danger. He howled, growled, and tried to rush at the window to warn off the intruders. After a month or so, when he varied his vocal range to include strange, high-volume staccato barks, and a noise very similar to a rooster crowing, he decided that anything with the temerity to pass the front windows had to be forcibly and noisily driven away. As well as our neighbours, this included birds, drifting leaves, passing bees, and cars he could only catch a glimpse of through the hedge. Amber and I had permanent headaches, and I suspect Skye did, too, as he would heave a sigh during each outburst, and retreat upstairs in search of peace and to lie on my bed.

Distraction methods didn't work. Even the prospect of cheese, that magical foodstuff, wasn't enough to persuade Charlie away from guarding duties. Thankfully, my neighbours were very tolerant, and laughed off my apologies; even so, the frequent sudden outbursts were wearing.

After all avenues had been tried and proved unsuccessful, I talked through ideas with Theo, who I will be eternally grateful to for many reasons, one of these being that she sent me a gift which saved our eardrums from perforating. A cardboard tube arrived in the post a few days later, shielding an innocuous-looking miracle worker: sheets of opaque film that stuck to the windows with static, letting in most of the light but blocking out what was going on outside.

Amber and I dampened the windows, applied the film, and waited for someone or something to pass by. It worked, and peace reigned once again. Three months later, when the hedge had grown quite a bit, I carefully peeled off the film, noticed how the extra light showed areas I'd forgotten to dust, and watched for signs of tension in Charlie. Thankfully, there were none.

Like all dogs, and especially fearful dogs, Charlie needs to know that I am his champion – ready and willing to protect him from situations he finds stressful or frightening, and to gently teach him, without putting pressure on him in any way, that he can overcome at least some of his fears. This also means understanding what he can't cope with, and ensuring that he's comfortable. Any signs of tension in his body and face – ears flat, a staring eye with the white showing (whale eye) – are signs that he feels anxious, which can play out in some odd ways at times.

... at the sight of me staggering across the bridge, knees buckling, with Charlie in my arms

Charlie has never drunk water from a bowl indoors, and will only sup from a bowl in one particular spot in the garden. He soon learned to use his food bowl indoors, and would keep an eye on me while I prepared his and Skye's meals, but, as with his water bowl, it has to be in exactly the same place every mealtime. If it gets moved an inch or two he slinks away, ears back and tail down, shooting worried glances at me, and refuses to eat. Even if he becomes over-enthusiastic while dining, and inadvertently moves the bowl himself, he'll swiftly back off until I return it to its rightful place.

The stream in the woods below our house

Charlie

is shallow enough in most parts for dogs to paddle safely, and is a favourite spot all year round. Skye likes to lie down and immerse himself in the shallows, with just his head above water. Charlie enjoyed walking through the woods but was scared of the stream for the first few months, and flattened to the ground if he thought we were getting too close. Then, one day, five months on, he followed Skye to the edge, dipped his front paws in, leaped backwards, then carefully crept towards the water, moving in slow motion with an expression of deep suspicion on his face. He reached down, dipped his head and drank. After that he was eager to test the water while Skye romped in it. One day, he leaped in after Skye and waded to the other side, leaving Amber and I open-mouthed with amazement and delight.

The walk through the woods takes us in a wide circle. We stroll down a tree-tunnelled path, over an old bridge across the stream by a small waterfall where we often see a Kingfisher, through a clearing and then through more woods. We pass a carved wooden statue of a hooded figure, half-hidden among the trees, and then cross the stream again over a beautiful bridge topped by a carved owl and moon, built by our friend and neighbour, James Wynne. This is the favoured dipping area for the dogs, but, for a long time, Charlie refused to walk over the bridge. He would take the first step onto the sloping ramp, look down, panic, flatten to the ground and refuse to budge, leaving no option but to carry him. By then, he had put on a significant amount of weight, and Amber would double over with laughter at the sight of me staggering across the bridge, knees buckling, with Charlie in my arms.

I wondered whether Charlie's depth perception was affected by having only one eye, making the combination of the ramp and the moving water below confusing to him. To test

this, the next time we walked through the woods, I suggested we bypass the carved bridge and cross the stream further along, over a rarely-used, rickety old bridge that had no ramp. Charlie didn't hesitate here, and happily trotted across.

A month or so later, after Charlie had negotiated the old bridge many times, I suggested we again try out a trip across the carved bridge. Amber and Skye went first, and I increased my pace so that we were close behind. Charlie crossed without a downward glance and received plenty of praise, followed by a paddle in the stream.

FLOATING HEADS

Another challenge to Charlie's perception was the four foot high fence along one side of my back garden. My neighbours, Cath and Phil, had not long bought Milly, a Labrador puppy, who was around four months old when Charlie first came to live with us. Charlie deeply disliked unfamiliar dogs at this time, and would hurl himself furiously at the fence each time he heard Milly playing in her garden.

To add to his heightened emotional state, the sight of my neighbours' upper halves over the fence sent him into a frenzy of howling, growling and barking. He was just getting used to the presence of people, but the sight of what appeared to be disembodied shoulders and heads moving along his territorial boundary was too much for him to cope with. Cath and Phil would speak to Charlie, whilst I stepped between him and the fence, rewarding him every time he paused to draw breath and check in on me, and one day Cath lifted little Milly to shoulder height so that Charlie could see her. This, of course, further confused and enraged him. He threw himself forward and upward, and only fast footwork as I blocked the way while Cath leaped back prevented little Milly from being badly bitten. As it was, Charlie managed to nip her nose. I suggested we wait a couple of days, and then take both dogs away from the gardens to work on some careful introductions, so that Charlie would become familiar with Milly, and, hopefully, less reactive. Cath agreed, and two days later we clipped on harnesses and leads, and went out into the road in front of the house.

It seemed that Duke knew, and was ready to leave

We started off with Cath and Milly way out in front, with Charlie and I following behind, which meant he could see her, and catch traces of her scent without being able to lunge directly at her.

At first, Charlie leapt around; pulled and twisted in an effort to escape his harness and get to Milly, making fearsome noises. We carried on walking, a safe distance apart, while I offered Charlie pieces of liver treats to create a positive association with Milly's presence. Eventually, after a few turns up and down the road, Charlie seemed calm, so we turned and began walking parallel to each other, with the dogs on the outside of us. Charlie lunged a few times, but then relaxed and looked at me to check whether I had any more treats. Gradually, we moved closer, and then switched places so that the dogs were walking side by side, with Charlie slightly behind Milly. He relaxed, wagged his tail, and moved forward to sniff her rear end. I kept a close watch for any stiffening of his body that would signal a lunge, kept his leash loose to avoid any possibility of tension transferring to him, and filled my palm with treats.

By the time we arrived at the small green area in front of our homes, Charlie was keen to make friends. We stopped on the grass and he play-bowed to Milly. Within moments they were having a great time romping around together, and Milly has been one of his best friends ever since. She's all grown up now, and they're the same height. When they hear each other while out in the garden, they jump up at the fence to peer over and say hello, tails wagging and tongues lolling, and Milly races over to play each time we meet out in the field.

SAFETY FIRST

Alongside keeping Charlie safe and at ease, I was also very aware that it was my responsibility to keep others safe around him, especially in view of the strong antipathy he initially expressed towards other dogs and certain people. Skye and Charlie developed a strong bond of friendship, and it was only later that this went through a very rocky patch. As a calm, gentle mentor dog to many other dogs, Skye had only ever experienced a rough time with two dogs who came to live with us. One of these was Duke, a beautiful, ten-year-old black Greyhound with flecked markings called 'ticking,' who came to us as an emergency foster after expressing his dislike of the toddler in his previous home. He was tall, even for a male Greyhound, and very easy-going much of the time.

Duke bonded strongly with me, and was always very sweet and affectionate towards me, but was very unpredictable with Amber, Skye, and everyone else. It turned out that poor Duke had a brain tumour, and his time was limited. He and Skye lived comfortably alongside each other for much of the time, but Skye soon learned to remove himself when he sensed Duke's mood had changed. Amber bore the brunt of Duke's aggression several times, and I had to step in to keep her from harm.

Charlie

When Duke's coordination slipped out of kilter and he found it hard to walk; when his aggression towards everyone but me intensified, and when he lost the ability to recognise anyone or understand where he was, Amelia came over to help him slip peacefully away. It seemed that Duke knew, and was ready to leave. Previously, he'd had to be muzzled while being examined, and this time, too, Amelia slipped a muzzle loosely in place. But Duke looked at her, climbed onto the sofa, offered his left front paw to Amelia, and rested his chin in my hands when I sat on the floor in front of him.

We removed the muzzle and Duke and I looked deep into each other's eyes while I told him how much I loved him. It was over in a moment, even before Amelia had finished the injection, and I sat with him for a while afterwards, shedding tears and stroking his noble head, before we wrapped him in his favourite blanket for his final journey.

Because my primary responsibility had to be towards Amber and Skye, I made a rule that I would foster dogs with any other issues except for active aggression towards them. This was their home, too, and they needed to feel safe here. Charlie's fear responses don't include aggression – he chooses to freeze or flee and it was only when he overcame many of his fears that a darker side emerged.

Unfortunately, one person whom Charlie took a dislike to right from the early days was Sam, Amber's boyfriend. Sam's a very sweet, caring young man, and I admired him for continuing to visit, considering he had to run the gauntlet between the front door and the stairs on each occasion. It must have put a strain on their relationship at times. Charlie had responded well to other men, so it was likely that Sam's slight nervousness of dogs brought out Charlie's bullying side – during the first sixteen months, Charlie always responded aggressively if he sensed fear or vulnerability. Despite Sam's attempts at friendship, Charlie would have none of it, and would redirect his hostility onto Amber, too, so a strategy was put in place to ensure that Sam could safely enter and exit the house.

Initially, I had to shut Charlie in the living room with me when Sam arrived and left, which caused a hangover effect as closed doors sent Charlie into a panic state that lasted for several days. We reconsidered how to better manage the situation, and found a new solution. Sam texted Amber when he parked his car, and Amber came through to let him in. I called Charlie and Skye into the kitchen, and rewarded them for coming to me, while Sam and Amber slipped quietly upstairs. It took a long time, but creating a positive association with Sam coming through the door eventually worked. Charlie has finally accepted Sam, and greets him with a wagging tail and friendly smile instead of full-on displays of aggression.

The real breakthrough came seventeen months after Charlie's arrival. Sam and Amber had been out for a bicycle ride together, and returned through the gate in the back garden. Charlie greeted them with enthusiasm alongside Skye, and trotted happily around the garden while Amber and Sam sat out there, enjoying the sunshine. Later, Sam came through to the kitchen to chat with me; something that hadn't been possible before that day. He lounged against the wall while we talked, and Charlie came in and out several times, looking relaxed and cheerful. Since then, Charlie has run to the front door to welcome Sam each time he visits, and often goes up to Amber's room to relax with them.

Seeing Charlie's attitude towards Sam

was yet another reminder that it was vital to be aware of the feelings and body language of everyone in Charlie's environment. As he is with me almost all the time, the need to monitor my own emotional state is crucial.

seven seLf-awareness

I am moved to tears the first time Charlie timidly clambers onto the sofa to lean in, touch his nose to my face, and curl up against me. He's been with us for around three months when he gathers the courage to climb up and join me, as he's watched Skye do many times, and his worried expression shows that he's not sure whether I'll welcome the extra company. I stroke his soft ears and he slides his head onto my lap, gives a soft groan of bliss, and falls asleep.

Dogs have evolved alongside us for at least 14,000 years; possibly much longer. During this time we have shaped them to fit in with our own designs in many ways, using selective breeding to change their size, appearance, and functions dramatically. Yet this is not a one-sided process, as dogs have contributed to our own species' evolutionary path. In their book *The Genius of Dogs: Discovering the Unique Intelligence of Man's Best Friend*, scientists Brian Hare and his wife, Vanessa Woods, suggest that it was wolves who adopted humans back in the distant past, rather than vice versa. The braver wolves frequented the refuse dumps created by early human communities, and those who were the least shy became scavengers. It's likely that some of the tamest, friendliest pups were taken into dwellings – and so the long process of domestication and collaboration began. We share a symbiotic relationship.

This affinity has been enhanced by genetic changes within dogs which enable them to 'tune in' to our micro-expressions and body language in order to interpret our feelings: an ability intended to keep them safe. If we are angry (and therefore potentially dangerous), we furrow our brows, narrow our eyes, our lips tighten, and our bodies tense. It makes sound survival sense for a dog who sees this to move away, rather than approach. Conversely, if our eyes crinkle with a smile, our lips curve upwards, and our bodies are relaxed and fluid, this invites proximity. Dogs can read us well – often better than we humans can read each other, and certainly better than many people read dogs.

We share a symbiotic relationship

Dogs observe us very closely, and are aware of the slightest change in our moods. They pick up subtle signals that we send out unconsciously, as well as consciously. Through the long co-evolutionary process, dogs have come to be accurate interpreters of our inner feelings. Recent studies of dogs' brains using functional MRI scans while they are fully awake have revealed increased activity in areas of the brain, especially the caudate, which shows that dogs understand and respond to our facial expressions and emotions. Neuroscientist Gregory Berns wrote a beautiful book called *How Dogs Love Us* about his ground-breaking research into this subject. Becoming fully conscious of our body language, and considering how dogs interpret this, can help us to transform our relationships with our dogs. This is a subject I wrote about in my book *The Heartbeat at Your Feet*.

The genes of feral dogs match those of domestic dogs, but their brains are wired differently. This system of hard-wiring takes place in all puppies, whether domestic or feral, starting in the womb, and extending through the first four months of life. If the mother is afraid of humans, this is transmitted to her pups, even before they are born. If a puppy has no contact with people during this formative time, he will be fearful of any that he meets later on.

We initially thought that Charlie was around eighteen months old when he was captured, as the estimated birth date on his pet passport is July 2011. However, long after Charlie came to live with us, an email conversation with Denisa Munteanu, his Romanian rescuer, revealed that he was at least a year, and possibly two or three years, older than we thought. This was borne out by Charlie's increasingly greying muzzle during 2014. Denisa told me that Charlie and his constant companion, Lenny, were first sighted in May 2011, and were adult dogs back then. She was contacted in September 2012 by a woman who had been leaving food out in the field that both dogs had made their home, and who had noticed that one of Charlie's eyes was seriously injured. Denisa drove to the area and took photos, then created an album to send to rescues along with a plea for help.

On January 18, 2013, Blind Dog Rescue UK stepped forward to foot the medical bills, and bring Charlie and Lenny to the UK, so Denisa captured both dogs and took them to a foster carer, where they lived in the garden and resisted attempts at contact from all except Denisa:

Charlie was very fearful but allowed her to touch him. Charlie's fear of humans was extreme, and any gesture was construed as threatening, but within a few months with us he began to learn to interpret the emotions of those around him.

COMMUNICATION BLOSSOMS

In the early stages of our relationship, Charlie struggled to understand the meanings of our facial expressions and body language. To him, everything was threatening and very scary, and it was immensely saddening at times to see him running back and forth in panic, or skittering off in fear, crawling away on his belly at the slightest sign of movement. Gradually, however, he learned this new silent language alongside the meanings of certain words, and it was exciting to watch the dawning of understanding shine out from his pale amber eye.

Love and empathy are natural emotions to feel in the presence of a dog

---

Dogs are adept at communicating with their bodies, and they're easy to read once we know how to do this. There is eloquence in their subtle gestures. The turn of an ear, the direction of a glance, and a shift in posture all speak volumes. Many of us are intuitively aware of what our dogs are telling us, even without studying canine communication, but delving deep into this subject and learning to 'speak dog' is both fascinating and exciting. It opens the portal to true communion with a being whose perspective on the world is very different to our own. When this occurs it is enriching for both us and the dogs we live with and encounter.

A great deal of our communication via our body language is unconscious or

Charlie

subconscious, though most of us manage to intuitively pick up the important signals. Dogs are expert anthropologists. Once Charlie learned to decipher our emotions and intentions, a new depth could be felt within our relationship.

Charlie observes me very closely, and I am constantly aware of his awareness. He misses nothing. Even when he appears to be fast asleep, the slightest change in my position prompts him to open his eye and check in on me. When I move, he moves. It's as if we are connected by an invisible thread that acts as a conduit for any slight shift in posture or emotion. He responds strongly to my inner feelings and silent signals, and his reactions have taught me to be acutely aware of what I am communicating through my movements and tone of voice. I've also had to learn to monitor my emotional states, because Charlie's reaction when I've felt upset, angry or sad was rather intense.

Love and empathy are natural emotions to feel in the presence of a dog who is frightened and confused about the strange new world he has been thrust into. Charlie has thrived through being loved and nurtured, and my heart has melted time and time again when he's come to lean against me, or nudge his head beneath my hand to ask for an ear rub. His expression of pure bliss when his invitation for physical contact is taken up is beautiful to see. When I'm feeling happy or excited he frolics alongside me, leaping up to offer his front paws so that we can dance together. During the first sixteen months, on the occasions when I was deeply upset, as when a dear friend passed away, Charlie's response was very different to Skye's gentle sympathy.

THE PRIMAL PREDATORY INSTINCT
Vulnerability – emotional or physical – brought out the predator in Charlie. Whereas Skye, and the other dogs I have lived with, have stepped in to watch over those who are sick or sad, Charlie made it clear that his intention was to eliminate the source of weakness in his social group. I've seen this happen numerous times when he's encountered people and dogs who are nervous of him: as soon as he sensed this, his instinct was to react aggressively. Because I could read the signs before he acted, I could keep those around him safe – but this primal instinct was also directed at me at times when I felt emotionally fragile during the first sixteen months he was with us. This meant that I had to be extremely self-aware, so that I could avoid giving out signals that would prompt Charlie's fierce aspect to come to the surface. The deep feelings of affection that he so obviously usually had for me were momentarily swept away. Trust was temporarily lost, and each time this occurred it took time and work on both sides for it to be rebuilt.

... and offered himself as mentor and friend

The difference in attitude between Skye and Charlie is clearly illustrated through their responses to vulnerable dogs. I've fostered a number of dogs, young and old, but during the two years prior to Charlie's arrival, all of the dogs who came to stay were seniors, sent to us by the Oldies Club, a wonderful charity that takes in dogs over the age of seven. I chose to foster some dogs who were terminally ill, because I have a very soft spot for old dogs, and feel strongly that their twilight time should be spent surrounded by love and comfort. Four dogs came, and subsequently passed away here, during the year before Charlie arrived. It was rewarding to know their final weeks or months had been happy, and also terribly painful experiencing the waves of grief that washed over us when each beloved friend died. Skye was gentle and caring with each of these souls. He encouraged them, lay by them, and offered himself as mentor and friend.

Shep, a fifteen-year-old, Collie/Husky mix, lived with a frail elderly lady before coming to live with us, and had received no exercise during the past twelve years. He was grossly overweight, the muscles on his back legs had atrophied through lack of use, he had severe arthritis, liver failure, agoraphobia, and canine compulsive disorder. Before he arrived I was told that he had only around two weeks to live, and my intention was to give him the best quality of life possible during that time. We all fell deeply in love with this dear old boy. He had the sweetest, sunniest nature, and Skye's influence made a huge difference to Shep's wellbeing and quality of life. Shep (or The Shepster, as he came to be widely known), adored Skye from the first moment of introduction. His tail assumed the characteristics of a helicopter propeller; his mouth widened in the most wonderful canine smile I've ever witnessed, and he desperately tried to stay upright, despite his back legs constantly letting him down, so that he could follow Skye around.

This sheer dogged determination to keep up with his agile new companion had the almost miraculous effect of getting Shep mobile again. It was a delight to watch him staggering around the garden and down the lane after Skye, and to watch Skye running over to wait patiently for Shep to get up again each time his legs gave way beneath him. With the combination of Skye's encouragement, Amelia's expert medical help in relieving Shep's pain and supporting his failing liver, and my behaviour therapy to help him overcome his fear of the outside world and

his obsessive compulsive issues, the years seemed to simply fall away from this lovely old dog's frame.

He lived a good life – the life I would wish for any elderly dog – with us, during his final three months, and we all missed him terribly when his liver finally gave up the fight. Shep will always live on in my heart and mind as a shining example of courage, persistence, and the ability to make the most of every moment.

The Shepster would not have fared well if Charlie had been with us, and I've accepted that my days of fostering elderly or traumatised dogs are over – at least for the time being. Charlie makes no bones about the fact that he despises weakness.

In May 2014, I had a visit from Emma, a friend who runs a dog training centre in Shropshire. She brought along Breeze, a very sweet, very nervous, eight-month-old Collie she was fostering at the time. Knowing Charlie's attitude towards nervous dogs, Emma and I took Breeze for a walk in the woods, and then kept her separate from my dogs. We divided our time between the space where my dogs were, and sitting out in the back garden with Breeze.

Skye and Charlie were aware of Breeze's presence, of course, and could smell her scent on both Emma and I. Skye kept looking for the new dog he could smell but not see, so after a couple of hours of keeping them apart we decided to try a careful introduction, with Charlie on-leash. Skye was delighted to make a new friend, and Charlie initially seemed eager to say hello. However, the moment Breeze cringed back and showed fear he lunged, snarling, and I immediately removed him before any hurt could occur.

I've discussed possible reasons for Charlie's aggressive response to vulnerability in others with my wolf expert friends, Toni Shelbourne and Isla Fishburn. It's likely that his instinct in the wild was to drive away those who were weaker than himself in order to increase his own chances of survival. Food resources would have been scarce, especially during the winter months, and Charlie was not one of the tamer dogs who would be comfortable scavenging in close proximity to humans. Denisa, his Romanian rescuer, is the epitome of patience and dedication. She looked for Charlie and his constant companion, Lenny, leaving food out for them until they were willing (or desperate enough) to approach while she stayed in the vicinity, which enabled their capture. Charlie's fear of humans and fierceness towards vulnerable dogs no doubt contributed powerfully to helping him stay alive.

Food resources would have been scarce, especially during the winter months

---

This element of Charlie's personality was to create considerable difficulties for several months, once he had overcome some of his fears and started to become more confident, seven months after coming to live with us. It resulted in a number of incidents that were uncomfortable at best, and downright unnerving at worst, but, fortunately, during the early bonding time, I was blissfully unaware of the additional challenges that were to come – and the even stronger bond that would be forged once we came out the other side.

My main priority was to help Charlie overcome his fears; relax, and enjoy life in a home, which meant I needed to be very aware of what Charlie's perception of my motives was likely to be in any given moment.

BODY-SPEAK

We sometimes use our bodies in ways that can easily disconcert nervous dogs. We throw our arms around when gesturing. We make silly faces when we're fooling around. We show our teeth when we laugh. We infringe on dogs' personal space by approaching too closely or too quickly. We make sudden, swift movements. We burst out laughing if something amuses us; when we're feeling good we may even hum a tune under our breath. In Charlie's presence, I had to ensure that I moved slowly and carefully at all times; that I kept my body language calm and clear, and that I monitored even my facial expressions as much as possible ... and humming was out for quite a while!

As so much of our mode of self-expression is unconscious and automatic, established through long habit, this proved to be an interesting exercise in self-awareness. I was used to moving carefully and fluidly because of the work I'd done previously with fearful dogs. Most, if not all of us, unconsciously 'pick up' on energy around others (a good example is that we can usually tell if an argument has just taken place in a room we enter), and dogs' senses are more powerful than ours, so I also used a visualisation method that helps to create a peaceful atmosphere. This involves imagining being surrounded by a bubble of calm energy, and it can make a remarkable difference to how we feel.

It certainly worked for me. Charlie began to display increasing signs of sociability: a soft eye, instead of hard stares; relaxed, more upright body language, with loose, floppy ears and a waggy tail, instead of slinking around; an open, curved, 'smiley' mouth. Best of all, he actively sought physical contact, sliding his body against me while I was standing, and nudging his head into my hand to invite gentle strokes. He began to approach familiar visitors, too, and would proudly sit beside them, his tail thumping against the floor.

We still had a long way to go on our journey together, but it was a delight to see signs that Charlie was beginning to really enjoy his new life.

VISIT HUBBLE AND HATTIE ON THE WEB: WWW.HUBBLEANDHATTIE.COM
WWW.HUBBLEANDHATTIE.BLOGSPOT.CO.UK
• DETAILS OF ALL BOOKS • SPECIAL OFFERS • NEWSLETTER • NEW BOOK NEWS

EIGHT Facing Fears

It's 2am, and I wake up in a state of terror, unable to breathe. My airway has suddenly closed off. I sit up and swing my legs to the floor, trying not to panic; trying to relax every muscle in my body while my heart beats in double time, and my lungs desperately strain to take in air.

Charlie is on his bed, close beside mine. He leaps up and skitters back and forth, aware that something is very, very wrong, then disappears into Amber's bedroom. A long two minutes later, as my vision darkens and I know I'm about to pass out, or worse, tiny amounts of air begin to filter through. The sound is horrible as I strive to draw in deeper breaths. Charlie runs back in, followed by Amber. She wears earplugs to bed because Charlie's morning yodelling wakes her far too early, and my feral boy, my hero, has persistently nudged and poked her to wake her, bring her to the awareness that an emergency situation is occurring, and I need her help. Charlie has very likely saved my life.

Fear is nature's way of ensuring our survival, and our bodies respond with a rush of chemicals that prompt us to freeze, flee or fight. The amygdala, an almond-shaped mass of neurons residing deep within the medial temporal lobe of the brain, sends an alarm signal to the hypothalamus, and a cocktail of adrenaline, cortisol, and norepinephrine floods our bodies. We experience this as a rapid heartbeat, raised blood pressure, accelerated breathing, dilated pupils, and sweating. If we exist in a state of constant fear this has a powerfully detrimental effect on our physical and mental health.

Although I did my best to alleviate Charlie's fear of absolutely everything during the weeks after he arrived, it was painful to see his confusion and panic over the slightest thing.

Everything in Charlie's new life was unfamiliar and very frightening to him; the things we and our home-bred dogs took for granted sparked a state of sheer terror in Charlie, and included: people; other dogs; rain; wind; thunder; fireworks; sudden movements or noises; the sound of the page of a book or magazine being turned; the scuff of my fingers against the mouse mat by my computer; the sound of the printer running and clothing brushing against furniture; the crackle of a crisp or cracker packet; the clink of cutlery; birds or leaves flying past; the sight and sound of cars; bicycles; motorbikes; children on scooters; music; the phone ringing; the television; any slight environmental change; any change of mood in people around him. The immense pressure of such continual, unalleviated stress would have pushed the strongest person into nervous meltdown, and, in retrospect, it astounds me that Charlie eventually learned to cope with most of these things.

Many of Charlie's fears concerning the intricacies of human life were easy to understand. Being in close proximity to people was something he'd not experienced until he was captured, and taken into Denisa's rescue centre. Denisa told me later that no-one except her could get anywhere near Charlie, and he was "very shy" and fearful, even with her. His only experience of transportation was the two-day van journey from Romania to England, followed by a long car journey to my home in Gina's care. Even now, after much careful work, Charlie will not go right up to a stationary car – the associated trauma that vehicles hold is still too great for him to overcome. Fortunately, Amelia visits him at our home when he needs any veterinary check-ups.

Fear is nature's way of ensuring our survival

Charlie's new world is necessarily a comparatively small place, bordered by the fields and woods at our home. I have counted my blessings many times – on Charlie's behalf as well as my own – that we live in the countryside, as each attempt to walk with him near a busy road has resulted in Charlie pressing himself to the ground, paralysed with terror. He and Skye have space to roam around, in an environment rich with the sights and scents of wildlife, and opportunities for socialising with their canine and human friends.

The range of a free-ranger

This limitation to one area is comforting to Charlie, and he has made it his new territory – a territory that he is happy to share with dogs he has befriended. Charlie's exercise area is similar in range to, and possibly much wider than, the distances he would have traversed in the wild. In a research study carried out by Thomas J Daniels and Marc Bekoff, culminating in a paper on spatial and temporal resource use, feral dogs and abandoned domestic dogs living in groups on the Navajo reservation in Arizona and New Mexico were captured, fitted with radio collars, and released. The animals were monitored through vehicular and hand-held antennae. In winter and early spring, the average range was 0.14 kilometres. In late spring/early summer, the range increased, averaging 1.62 kilometres. The furthest range, of 5.89 kilometres, was traversed by one of the females.

Charlie likes to follow a familiar path, and I'm sure this would have been his routine in the wild, too. He's a creature of habit, and although he'll walk alongside me if I take a different entry point into the woods, his eye is focused on the usual path, and his preferred route is always the same.

This takes us through the top entrance of the field, which gives a view of distant hills; past a small, fenced-in children's play area, along a wall that backs onto the field, then down towards the woods. Three paths lead steeply down through the trees, but Charlie always chooses the path closest to the edge of the field, sheltered on one side by crops. He follows the path down, almost to the end, turns right to cross an old bridge that leads over the stream, has a drink, and sniffs around in the clearing while Skye paddles into deeper water; then trots happily through the trees, over the stream again by a different bridge, and up the steep bank that leads back to the bottom of the field. From here, he turns and follows the path that runs along the back of a few houses, past the horses' field, and in through our back gate. If the gate has swung shut in the wind, he waits quietly for me to open it. I'm sure he could do this habitual walk in his sleep.

The HULK becomes a friend

Charlie perceives anything unknown to be potentially dangerous, and his reactions are

Charlie

lightning-fast. With unfamiliar sounds and objects, his default response is to freeze and flatten to the ground, or flee. With unfamiliar people and animals outside our home, his instinctive response during the early months was to lunge. Twice daily we went to the field, and each time we met at least one friend and neighbour. Skye, who is a friendly soul, is always thrilled at the prospect of a kind word and a stroke, whereas Charlie transformed into the canine equivalent of The Hulk: standing tall, hackles raised all the way down his back as well as on his shoulders, lunging, barking and snarling ferociously. It was a sight to make even my most enthusiastically dog-loving friends take a well-advised step back.

When this happened, I asked everyone I saw to stand well back, out of range, and stay very still while I waited for Charlie to pause and draw breath before I spoke his name. Once I had his attention, his favourite liver treats were offered to him, along with a single calm invitation to sit. Eventually, Charlie would realise that no treats were forthcoming until he sat, and a seated dog is in no position to lunge: in behaviour terms we call this the teaching of incompatible behaviours. Each time he jumped back up to start again, I caught his attention and repeated the request.

My neighbours were wonderful. Fearsome though Charlie's attitude towards them was at first, they expressed a willingness to help me teach him that there was no need to defend himself. Within two weeks of practicing, Charlie had reached the point where he would sit beside me while I briefly paused to say hello to everyone, before moving on into the field. After a month, he was wagging his tail at the sight of people he recognised, and sitting without being asked the moment they approached. It took a little longer with unfamiliar people, but now Charlie steps forward to greet everyone we see, known or unknown, if they speak to him, and he has several real favourites whom he bounds towards with an expression of pure delight, wriggling and bouncing in his eagerness to say hello.

Positive reinforcement is a system of rewarding the behaviour we wish to encourage in our dogs, and want them to repeat. It works just as well with people – I used it on my children while they were growing up, too. Dogs soon learn which behaviours result in their receiving something that makes them feel good, be this a food treat, praise, a stroke or a game – whichever the dog likes best at the time.

Charlie has gone hungry in the past, and, although he was underweight when he arrived (I could see his ribs and hip bones beneath his thick coat), he was nowhere near emaciated, as some of my cruelty case, fostered and adopted dogs have been, so he had clearly done well with fending for himself in the wild. Nevertheless, food – the smellier the better – was guaranteed to catch Charlie's attention, and proved to be the ideal motivator.

Another thing Charlie loves to do is run, and he's extremely fast on his paws. This proved useful when we encountered a car, bicycle, motorbike, or a child on a scooter, which, stationary or moving, would send Charlie into a state of panic. His natural response was to flee, so I would call his name in a 'happy' tone of voice before he had chance to freak out, and jog swiftly past whatever was the cause of his fear. This enabled him to follow his natural instinct: to safely act on, and release, the cocktail of stress chemicals flooding his system. It became a game, and before long I could jog more slowly past, then walk quickly, then simply walk, and Charlie remained calm. Entering a car is a different matter, and this is still a work in progress, but as there's been no requirement, yet, for Charlie to be driven anywhere, I can take this slowly without subjecting him to any pressure.

... and burst forth in a frenzy of barking and lunging

The first time Charlie saw a cat, he really didn't know what to make of it, so I could only assume he'd never encountered one before. Initially, he stood very still, looking puzzled; then looked at Skye to gauge his friend's reaction. Skye, being a Deerhound/Greyhound/Saluki mix, bred from a long line of working dogs, thinks cats are for chasing (fortunately, this prey drive doesn't include small dogs!), so he leapt in the air and prepared for action. Wearing his harness and lead curtailed any further hunting activity, but Charlie immediately took his cue from Skye and burst forth in a frenzy of barking and lunging. The cat stalked off, looking superior, and I waited for peace to return before moving on. The village cats are very dog savvy, and one of them, a beautiful Burmese, used to stalk the dogs, and pounce on them from within the bushes. Skye was caught out by her one day, and now has a healthy desire to leave the cats to their own devices ... Charlie's swiftly learning to follow suit.

When Rabbit the pony moved into the field just below our garden, Charlie was completely flummoxed. The combination of a large, unfamiliar creature just beyond his territory, and much to-ing and fro-ing of people going down to settle her in and visit her, resulted in Charlie making a great deal of noise to alert me to their presence, and warn them off. The first time he saw Rabbit at close quarters as we passed the field on our way home from the woods, he scuttered across to the other side of the lane and dived into the hedge. Later on, Rabbit was joined by Black Velvet, a young

horse, but, by then, Charlie had grown used to seeing an equine head peering at us over the gate, and he adjusted far more quickly. It was a proud day when he walked past them without reacting, other than to wag his tail at their owner.

SHIFTING LANDMARKS

In the wild, any environmental change can indicate potential danger. The presence of something in a previously empty space could be a predator. The absence of a familiar landmark can cause confusion. Charlie needs to feel that his world is uniform; always the same. He doesn't cope well with change, and when I've needed to move pieces of furniture temporarily he's reverted to his old, fearful self – anxiously running back and forth, belly to the ground.

For a couple of years I had a rug on the kitchen floor, whose purpose it was to soak up some of the mud the dogs track in from the garden, because their pawprint works of art involve a lot of cleaning up, especially in wet weather. The rug, once rather pretty, became so tattered that I moved it out into the passageway by the back door. Charlie trotted through to the kitchen, clearly intending to carry out the next phase of his den-digging endeavours in the garden, and stopped dead with cartoon-character speed, his body leaning back from his extended front legs, and his face a mask of absolute horror. Slowly, he reversed, turned, and then bolted into the living room, refusing to come back through. Nothing would induce him to put his paws on the kitchen floor, not even an offer of cheese. Charlie was adamant: the rug had to stay. On that day I relented and brought it back into the kitchen.

A plan was needed to help him accept the change. Twice more I moved the rug, putting it back an hour later without making a fuss of

Charlie

Charlie, to show him that some things could migrate without causing him harm, and that no pressure would be put on him. The second time, he peeked anxiously through the doorway but didn't bolt, so the following day I again moved the rug and left it in the passageway.

Charlie peeped through the doorway a few times, then skittered right through the kitchen and out into the garden. It took him a while to gather the courage to come back indoors, but after negotiating it once, he happily trotted in and out without turning a hair.

Charlie's reaction to rain, as well as to storms, was extreme

---

It was surprising to discover that a free-ranging dog could dislike rain so much, considering he had lived out in the elements for so long. Many dogs are afraid of thunder, but Charlie's reaction to rain, as well as to storms, was extreme. He would repeatedly run back and forth in a state of abject panic, belly close to the ground, ears pinned tight against his skull, his eye so wide that the sclera – the white area around the iris – was clearly visible. I could only imagine how frightened he must have been out in the open in rough weather. The sound of rain is intensified indoors because it comes down part of the chimney, making plinking noises as it hits the shelved area partway down, and this added to Charlie's distress. It was almost fourteen months before Charlie could sleep peacefully through rain.

Thunder was the catalyst for a huge advance in Charlie's exploration of our home. During the first six months he steered clear of certain areas – the utility room beside the kitchen; Amber's studio in the garden, and the stairs. As Charlie had always completely ignored the stairs,

not even glancing at them if Skye climbed a few steps for a better view of visitors coming through the front door, his obliviousness to them made me think about what is known as perceptual blindness. This is not a vision defect – although Charlie has only one eye, his vision in that is very acute – but a state in which the observer, faced by something that he or she has no reference for, simply blanks out, and doesn't perceive as being there at all. Charlie never even glanced at the stairs, although he passed them every time he approached the front door. If I spoke to him from the staircase he was unresponsive; it seemed as if he had no inkling of my presence.

One evening there was a tremendous thunderstorm. Skye went up to lie on my bed, his favourite place of comfort and safety, but Charlie was utterly panic-stricken. He raced back and forth, in and out of the living room, coming to me occasionally to press himself against me for a moment, trembling all over, before running off again. The clicking of his claws on the wood floors merged with the thunder and the sound of torrential rain. Suddenly, I could no longer hear him. I went to check, and saw him halfway up the stairs, frozen in place, too scared to go further up, and with no idea how to turn around and come down.

Carrying Charlie back downstairs seemed to be the only option, though he was too heavy for me to lift easily by then. As I moved towards him, Skye appeared at the top of the stairs, paused briefly to assess the situation, and ran down the stairs past Charlie, looking pointedly at him as he passed, as if to say "This is how it's done." At the bottom he turned and ran back up, then waited at the top, his attention fixed on Charlie. As Charlie remained in place, Skye demonstrated again, running down, then back up and waiting at the top. Charlie raised his head, focused on Skye, and ran up the stairs and straight into Amber's bedroom, making himself comfortable on the dog bed in there. He stayed there for several hours, and after that night he went up to sleep in her room every evening for the next three months.

During that time Charlie ignored the doorways to my bedroom and the bathroom, just as he had previously ignored the stairs, until one night, as I was preparing for bed, he peeked in and took a flying leap from the doorway onto the dog bed in my room. Since then he's chosen to sleep in my room every night.

Learning Through Imitation

Humans aren't the only species to learn through imitation, and studies with dogs have revealed that they learn from each other, and from us through mimicry. Stanley Coren, professor of psychology at Columbia University, explains how dogs learn from observing other dogs in his article for *Psychology Today* dated January 2013. Skye's tutorial in how to negotiate the stairs worked far better for Charlie than if I had tried to coax him up or down, and was yet another demonstration of Skye's excellent mentoring skills.

Behavioural ethologists at Eötvös Loránd University in Budapest, József Topál and Ádám Miklósi, devised a method that they called 'Do as I do,' and discovered that dogs can learn to imitate human actions, even after a time delay. This is one of the principles that I tapped into when teaching Charlie to be comfortable in his new environment.

Dogs take their cues from us. If we're anxious when another dog approaches, our dog may think there is real cause for worry, and is more likely to become reactive. If we're relaxed, we communicate to our dog that there is nothing to fear. So, with Charlie's many fears, it was vital that I showed him, through my own body language, facial expressions and responses, that he was perfectly safe.

Charlie's reaction of fear and aggression towards the television the first time I turned it on was allayed when he understood that I wasn't worried. Other fears took longer to overcome, and some of the longer-standing ones were due to his inability to process unfamiliar sounds.

I read at every opportunity; whenever I have a few minutes free, I pick up a book. What, to me, was the pleasant sound of pages being turned sent Charlie skittering out of the room each time. On returning, he would go to his bed and lie down, his face and body tense, staring hard at me as he tried to figure out why I was making such a peculiar noise. It was almost a year before he could relax and fall asleep while I was reading. The rustle of a crisp or cracker packet had the same effect, though this was negotiated much faster after I took the packet through to the kitchen afterwards, then called him to follow me, and gave him a morsel to try. Nowadays, he gets a distinct gleam in his eye when he hears that noise. The sound of the printer churning out pages still prompts Charlie to howl at full volume (I made a video of this that can be seen on YouTube – https://www.youtube.com/watch?v=Awmq90mhnTs).

My feral friend needed to be taught to accept the presence of other dogs without feeling threatened, and his adventures in socialising are described in chapter 10. As far as Charlie's concerned, once a friendship has formed he is completely devoted, and his loyalty is deeply touching. Kelsey, a beautiful Golden Retriever, was already good friends with Skye, and Charlie was soon smitten by her easy charm and relaxed manner. One day, as Amber and I were heading into the field, Kelsey ran out of her house and

Charlie

came dashing over to join us. A reunion dance was performed between her and my boys, play-bows were engaged in, and there was much joyous leaping about. As Kelsey's carer, Felicity, had no idea that her fun-loving girl had gone walkabout, we took her home and were invited in so that the dogs could play together in her back garden for the morning.

Once on the grass he was fine, and had a wonderful time playing

---

Charlie was in full social mode until he realised that I expected him to step over the threshold of an unfamiliar house with me. Immediately, he flattened to the floor and refused to budge. As Skye had followed Kelsey in, and Felicity was waiting for me to join them, I had no alternative but to pick up Charlie and stagger into the house.

Once set down on the living room carpet, Charlie went straight back into 'freeze on the floor' mode. For a moment, I debated whether to pick him up and leave, or pick him up and carry him into the garden. Through the window I could see Amber watching Skye and Kelsey racing around, tails whirling in circles of joy, so I carried Charlie out to join them. Once on the grass he was fine, and had a wonderful time playing – until it was time to leave. As he refused to re-enter the house, I carried him back through, and he happily trotted home.

Charlie is not a 'visiting' kind of guy, though he welcomes visitors – and he has grown far too sturdy to be comfortably hefted in my arms!

NINE Patience

I've been out for most of the day on a vet referral case, working with a dog with serious aggression issues. This is the longest I have left Charlie in Amber's care, and I wonder whether he will regress to skittering around while I'm away. He's still going through his phase of avoiding doorways and bolting if anyone is in the hall, and he depends heavily on me to provide a sense of security.

Skye and Charlie are in the back garden when I return, and Skye runs in to greet me when he hears me talking with Amber. Charlie trots into the kitchen a few moments later, and suddenly realises I'm back. There's a rapid scrabble of paws as he accelerates and races into the living room where I'm crouched on the floor, greeting Skye. He leaps into my arms, knocking me off-balance, and his entire body wriggles as he rubs himself ecstatically against me, covering me with kisses. I kiss his soft ears, heart swelling, as he joyfully pushes his face against mine.

The process of creating a bond with a fearful dog – whether he or she has experienced abuse in the past, or has been rescued from a puppy mill, and has had no opportunity to learn that humans can be kind – is a slow one. Patience, understanding, kindness and consistency are the keys to gaining that vital trust. Charlie has constantly reminded me that all good things come eventually, if only we are willing to put in the work, and wait.

Patience is an easy quality to access when there's no pressure and all is running smoothly. It's a little more challenging when the odds seem to be stacked against us, work deadlines are looming, and life descends into a temporary state of chaos.

It's absolutely necessary to take slow, careful steps, and avoid the use of any pressure whatsoever with frightened, nervous and anxious dogs. This doesn't come naturally in our hectic society, where everything in life rockets along at a fast pace, and a multitude of tasks need to be compressed to fit into tight schedules. I understood that it was vital to hold back, slow down, and eliminate all expectations. During Charlie's most testing times it often felt like an achievement to fall into bed at night, simply knowing that no human or non-human animal had come to any emotional or physical harm that day, and everyone's eardrums remained intact. Those days were exhausting, but they were shot through with bright moments, and there was a certain joy in knowing that the duration of each setback was shorter every time. "All things must pass" became my motto for quite a while.

Because many of Charlie's regressions occurred just as life seemed to be slipping into a smooth, easy pattern, there was never any possibility of becoming complacent. These setbacks were linked with any form of change, however minor. As they were also often sparked by upsetting news or situations from outside our

Charlie

family group, after a while, a clear link between my emotional states and Charlie's behaviour emerged. The invisible thread that connected us became like an umbilical cord: transferring the nutrients of love and affection, but also the toxins of distress and negativity. This emphasised the need for me to monitor my emotions and the causes of these, which meant being more careful about what and who I invited into my life. It proved a good code to live by.

... a clear link between my emotional states and Charlie's behaviour emerged

---

Charlie's habit of howling, and suddenly giving voice to his strange, crowing bark, both in the house and in the garden, was hard on the ears, and made concentration nigh impossible. Although the sound of his bark was so peculiar that it was actually comic, Charlie needed to learn that it wasn't acceptable to yodel continually. Placing a static film over the living room windows shut out the stimulus of any passers-by, which helped Charlie to be more relaxed indoors, and by the time I took this down, the garden hedge had grown enough to block out all but the head of the occasional tall person. This worked beautifully. Outside was another matter ...

Each time Charlie went into his full-on vocal repertoire in the back garden I brought him inside. For a while, when his much-needed weight gain made it hard to carry him, I kept a leash in the kitchen to clip onto his harness or collar after catching him. This inevitably sparked off a game in which Charlie raced around the garden, twisting and turning, while I dodged and counter-dodged until I managed to intercept him. After months of these impromptu dance routines, Charlie's passion for cheese provided

the perfect enticement. I could stand inside the kitchen, call him, and he would dash through and wait expectantly while I quickly shut the door and rewarded him (and Skye, who was never far behind) with a morsel. Gradually, life became quieter.

The Charlie Rules of moving slowly and speaking softly stayed in place for almost a year before they could be gradually relaxed. Sudden movements towards him, especially from people he doesn't know well, still prompt a startle, cringe or bolt response, though I can run to answer the phone or door nowadays, though at the risk of tripping over Charlie, who joins me in the race towards the source of the noise!

The appearance of unknown dogs brings out the tough guy in Charlie, and this necessitates a quick visual scan of the field when we enter. His tail flags his delight when he sees a canine friend, but the sight of an unfamiliar dog prompts his hackles to rise, while he engages in full-throttle shouting and lunging. Socialising Charlie was a slow, careful process, and my friends in the village were invaluable in helping me do this, as you'll read in the next chapter, but he now has a number of dog friends with whom he loves to play. His bright eye, strutting gait, and air of self-satisfaction after a play session are heart-warming to see.

Charlie's brief but hair-raising career as an escape artist was testimony to the importance of teaching reliable recall. Fortunately, this happened after he had learned to come inside from his barking episodes, so by then he associated my calling his name with something good coming his way.

One morning, we had just entered the field when I realised that two friends were over on one side, chatting, while their dogs stood beside them. Toes, a beautiful, fifteen-year-old

Collie/Retriever mix, had a deep dislike of male dogs he didn't know well, and it had taken him almost two years to accept Skye's presence. I was immensely fond of him, and the feeling was mutual – each time I visited his home he would lie on my feet and nudge me for strokes, and this more than likely helped avert a catastrophe. Catching sight of us, Lily, a tiny Manchester Terrier, rocketed towards us, flirtatiously dashing back and forth in front of Charlie and Skye. One moment, Charlie was beside me; the next, he had performed an impressive double somersault in the blink of an eye – and was gone, leaving me holding a leash and empty harness.

Lily, suddenly realising that Charlie was no longer safely attached to me, raced back to her carer, with Charlie in swift pursuit. This meant she was headed right towards Toes, and so, by default, was Charlie. My heart skipped a few beats with the realisation that blood may be shed, and no way could I cross the field before Charlie reached Lily and Toes.

... an impressive double somersault in the blink of an eye - and was gone ...

Instinctively, I called Charlie's name, using the 'happy' tone he was accustomed to. To my amazement and pride, Charlie turned away just a few inches from contact with Toes, and ran back towards me. I crouched on the ground, my arms held wide, and caught him as he hurtled into me. Skye had been calmly waiting beside Amber, who quickly slipped Charlie's harness over his head so that I could clip it in place. As Murphy's Law would have it, that was one of the few days I'd forgotten to bring the treat bag, but Charlie seemed happy enough with being fulsomely praised.

Giving Charlie the ability to make choices, and rewarding the behaviours that I wish to encourage, have helped him to adjust to his new life without putting him through the strain of additional trauma. I've always been strongly against the use of force: it's unpleasant and unnecessary, and the heightened stress and anxiety it creates inhibits the ability to learn. Force methods destroy trust, so are counter-productive to developing a harmonious relationship, and the bonds we develop with our dogs must have trust as the foundation. A calm, patient approach is needed with any dog, and with Charlie it was vital. It's helped me to find ways to resolve Charlie's issues, and has accelerated his recovery after each setback.

As the slightest change always sparks a regression, it's likely that setbacks will always occur, and each time it's important to go back to the beginning with him – to patiently rebuild his trust and confidence, and help him to cope again. An unusual number of visitors, even though he's happy to see them; me being out of the house more often than usual; the sight of an unfamiliar object, or any small change in his environment, all worry Charlie deeply. He reverts to skittering around, belly to the floor, cringing if anyone glances in his direction. As this stage of dis-ease begins to pass, he becomes over-excitable, and will jump up to grab at me each time I move. Charlie uses his paws like a cat, curling his toes inwards and spreading them so that his nails are extended, and my arms bear a number of scars where he's accidentally raked them when he's clutched at me. Yet each time he gradually readjusts and relaxes again, and the time frame between his regressions grows longer.

Watching Charlie make that shift from being a terrified, confused wild creature to a happy, mischievous furry friend has been immensely rewarding. One of the many side-effects of Charlie's transformation has been his ability and desire to bond with people beyond our small family group.

Charlie the rescue advocate

Two close friends came to visit from America in October 2013. Kac Young and Marlene Morris are long-standing friends whose abundant motivation, creativity and productivity inspire me greatly. Their accomplishments are awe-inspiring. They are both authors, with many books published; Kac is a former television producer, and Marlene runs a ministry. In all the years I have known them they have loved dogs, but have been very much 'cat people.'

Charlie was to change their lives, as well as mine. Kac and Marlene have known Skye since he was a puppy, and have a deep affection for him. However, Charlie's response when he met them was extraordinary. He flirted, he offered paws to hold, he sat close, snuggling up to them. He completely captured their hearts.

Not long after they returned home I received an email with some photos of Talulah, a small dog they had met and fallen in love with at their local rescue shelter. She settled in wonderfully, though Kac and Marlene informed me that their cats were no longer speaking to me. Recently, another photo arrived. Talulah now has a companion, a young male called Truffle, and the cats, after a long period of haughtily retreating to a different area of the house, decided they may as well venture forth and accept that the invaders were there to stay. In working his magic on my friends, Charlie helped to save the lives of two more dogs.

The wild aspect of the self can wait. It has boundless patience. The process of stalking is slow and careful, be it employed by a hungry feral dog to secure the next meal, or in the achievement of a goal for us humans. Careful steps and waiting for the right moment are essential: in the wild, in particular, any impulse to rush in could result in loss instead of gain.

Like all dogs, Charlie takes each moment as it comes. He doesn't think about what's around the corner, unless a cue, such as the appearance of a meal or his leash, is offered to him. He sees what needs to be done – whether this involves action or inaction, a chase or a nap – and this bestows an intense purity to each of his moments. He focuses simply on what is there, right in front of him, without considering what could possibly result from this in a week or a year. Charlie has an instinctive ability to experience the full spectrum of life (its good and bad times) without the human trait of expectation. Through observing this in him, I have learned to step back a little, to wait, and to be more aware of the implications of my actions in this moment.

Ten MakING FrIenDS; avoIDING enemies

It's early evening, and we've been down in the woods, Amber and I chat whilst Skye and Charlie have a good sniff around and a paddle in the stream. As we step out from beneath the canopy of trees I see a small blur of pale fur racing down from the top of the field. Skye's head goes up and he whines with excitement. Charlie follows Skye's gaze and dances on the spot before they race together to meet little Millie, a young Cavachon, as she bounds towards us. Paul, Millie's carer, follows on as the dogs meet halfway; it reminds me of those movie scenes where the long-lost lovers run towards each other, arms outstretched, rushing headlong into an embrace.

Millie looks tiny beside Skye and Charlie, but she leaps in between them and they greet each other ecstatically. She stands on her hind legs to lick their muzzles, and then briefly rolls over, offering her belly for them both to sniff, before jumping up to join in a madcap circular game of tag. We stand and watch, all of us uplifted by the joyful reunion playing out before our eyes.

Millie was recently attacked by two Labradors. She's been nervous of entering the field since then, and Paul is happy that Skye and Charlie have reminded her that she has friends to play with whom she can trust. It is a profoundly beautiful, healing moment.

For a while, Charlie needed a great deal of help to learn to accept the presence of any dog apart from Skye. Once he began to settle in with us he became very territorial, and would patrol the hedge in the back garden, warning off any dogs who had the temerity to pass by on their way to the woods. This would have been his natural instinct in the wild, and he made it abundantly clear that he would defend his territory against all comers.

He now enjoys greeting everyone he sees during our walks

When I started to take Charlie for walks, he demonstrated fear-aggression towards people, and an intense dislike of other dogs. I am fortunate to have good friends in the village who were willing to help me socialise him. They have befriended all of the dogs who passed through my home for either temporary foster care before being rehomed, or for terminal care, and I've become friends with their dogs, and helped some of them to overcome behaviour issues.

Charlie's reaction to an approaching person was to lunge, growl, bark and snarl, and in Chapter 8 I described how I taught him to enjoy meeting new people. Liver treats, cheese and praise were his rewards for staying back and focusing on me: then, as now, I was always careful to let Charlie see that I'm his protector, and will keep him from harm. This worked so well that he now enjoys greeting everyone he sees during our walks. He's become very friendly with most people, though (like many dogs) will back away if a stranger impolitely approaches at speed and looms over him.

Teaching Charlie acceptable manners around other dogs took a little longer. His reaction towards any that he saw was extremely aggressive, so I spoke to Pete and Angie, whose liver and white Springer Spaniel, Mayzie, is very confident and friendly. I've known Mayzie since she was a puppy, and she's a typical Springer – happy-go-lucky, energetic, and very sociable. Pete and Angie were willing for Mayzie to be involved in Charlie's training, and I reassured them that I would be in full control, and that their sweet girl would be safe, however reactive Charlie was during careful introductions.

Mayzie is usually off-leash while out, but Pete clipped a leash on for her first few meetings with Charlie. I didn't want to risk her coming too close to him and possibly being bitten. Filling my pockets with liver treats, I took Charlie out through the back gate to the top of the lane where it widens and meets the small road that leads to our house. Pete was there, on the grassy area, with Mayzie, who wagged her tail and moved to greet me.

Charlie exploded into full-on barking, growling and lunging, his hackles right up along his neck, shoulders and back. I asked Pete to walk ahead of us so that Mayzie was facing away from Charlie, while I at first moved further away and got Charlie's attention before turning around to follow them. Charlie exploded again, so we repeated this manoeuvre over and over, moving away each time, until he was calm. By the end of the session we could move to within three feet of Pete and Mayzie without Charlie reacting, but I knew it would put too much pressure on both

dogs if we narrowed the gap any further. We ended the session.

During the next meeting Charlie was much calmer until we reached the critical three foot gap, which was too close for his comfort. I moved back, just beyond this, and he was willing to wait beside me while I fed him treats and spoke quietly to him. Mayzie was wonderful, casting glances across in the hope of getting her usual greeting from me, and accepting the situation when it was clear I wasn't coming any closer.

The third meeting occurred spontaneously.

We were heading home from a walk when Pete and Mayzie happened to pass us on their way up the road. Skye danced forward to greet her, reaching her first, and they engaged in their usual enthusiastic welcome. Charlie followed Skye's lead and stepped towards her, tail wagging, so I kept his leash very loose, prepared to move him away at the slightest sign of tension. To our surprise he sniffed her and then play-bowed, and they played together.

Mayzie has a very distinctive, high-pitched bark, so we can always tell when she's out in the field. Once Charlie had made friends with her, he would go into the garden and howl as soon as he heard her in the distance. Mayzie pokes her nose under the back gate to say hello as she passes by, and has had quite a few play-dates with my boys in the garden.

As soon as Gunner saw us he raced over, tail wagging

We used the field for introductions with other dogs, and Charlie's circle of friends has grown. As well as Mayzie, there's Cyder, a male Cocker Spaniel; Milly Labrador, who lives next door; little Millie the Cavachon; Rosie and

Connie, both Fox Terriers; Poppy the Border Collie; Kelsey the Golden Retriever; Bear the Miniature Schnauzer, and, most recently, Gunner, a Cocker Spaniel puppy.

The first meeting between Charlie and Gunner was impromptu, which demonstrated very clearly how far Charlie had progressed. We were just coming into the field as our friend, Geoff, was approaching the entrance, on his way home with Gunner. The puppy knew me well, because I'd visited Geoff and his lovely wife, Shirley, numerous times since he arrived, and two groups of my students had helped with Gunner's training during practical study days with me. As soon as Gunner saw us he raced over, tail wagging.

Charlie immediately went into a play-bow

I looked at Charlie for signs that he wasn't comfortable about being approached, but his face and body were relaxed and loose. Geoff, having seen Charlie at his worst in the past, called out "Is he okay?" and I confirmed that all was fine just as Gunner reached us. Charlie immediately went into a play-bow, and he and Gunner had a romp on the grass while Skye stood to one side after a quick, friendly greeting, watching the fun.

Just as some humans are more social than others, not all of the village dogs are friendly, and several are seriously dog-aggressive. It's easy to tell when they're on their way to the woods, because some of them noisily try to dig under our back gate to get to Charlie, while he goes into full defensive mode, and he and Skye patrol their side of the hedge. We've had a few hairy moments when we've turned the sharp corner in the lane, come face-to-face with one or another Nemesis, and had to either turn back into the field, or move swiftly through the gate to get our dogs safely into the garden in time to prevent a fight breaking out. However, Charlie now has far more friends than adversaries, and he's become increasingly keen to meet and greet new dogs.

SMALL PEOPLE

Another aspect of Charlie's social life was learning to cope with the presence of children. As well as the small group of children living in the village, my friends' grandchildren visit, and most of them pass our garden to go and play in the field and woods. Their high-pitched voices frightened Charlie during his first months with us, and sent him into a paroxysm of barking every time he heard them. He would have had no contact with children in his previous free-ranging life, so it wasn't at all surprising that he was so unnerved by them.

Children often pass us during our walks, and, as news of Charlie's unusual past and his gradual progress in adjusting to home life had created a lot of interest in the village, everyone knew that he needed to learn to feel safe from a distance at first. The children would walk by a few feet away, without stopping, and Charlie received a reward and praise each time he saw them. After a couple of months he was stepping towards them, wagging his tail, instead of barking, cringing, flattening to the ground or hiding behind me.

The next stage was allowing the children to step forward to say hello to Skye, who is very fond of small people, while avoiding getting too close to Charlie. From there, seeing that Charlie felt comfortable with this, they spoke to him without touching him. Before long he was wagging his tail when he saw them, and moving forward to touch their hands with his nose. The day our friends' grandchildren came briefly into the back garden and Charlie skittered around for a few minutes, then followed Skye over to greet them politely, was a very proud day for all of us. He now has favourite small people who he greets with a joyful dance and a lick of their hands.

The key to socialising Charlie was to take great care to avoid letting him become stressed or overwhelmed. I used this softly, softly approach consistently, and it paid off. The dawning understanding that he could trust me to take care of him and keep him safe helped Charlie cope with new situations and experiences. However, a major change in Charlie's new life came, by necessity, without the usual slow, gradual introductions: the arrival of a foster dog.

A new companion

Foster dogs are always an unknown quantity. Some are assessed in rescue shelters before being moved into foster care; some come from homes where their carers have passed away, or can no longer keep them for various reasons, and many come from dog pounds, where their background is a mystery. Even a dog who has been assessed can react very differently in a new environment, and although rescues try to ensure that the match will work with resident dogs in the foster home, there's no way of knowing whether they will all get on well.

I've been fortunate in that Skye is a remarkably tolerant dog who accepts all comers, and does his best to help them feel comfortable around him. I've fostered many dogs, and all but two – who turned out to have severe dog-aggression issues in subsequent homes – settled in beautifully with my own dogs. As fostering has always been important to me, I took in a beautiful, Whippet-type dog called Saffy once Charlie seemed stable in his new life. The rescue and I discussed beforehand what impact

this may have on Charlie, and my main concern was that he wouldn't accept the presence of an unfamiliar dog within his territory. My only rule around fostering is that the new dog must get on harmoniously with my dogs, as it would be unfair and cruel for my dogs to feel unsafe in their home, their place of security, so we agreed that if there was serious conflict, Saffy would have to be moved to a different foster home. The impact of having her here with us was far different to anything we had considered.

It was beautiful to watch them chasing each other

Saffy's story was a sad one. She was a Greek street dog, and was in a terrible state when taken into rescue. She had been de-gloved – all the flesh on her paw pads was torn away from being dragged along a road, most likely behind a car or truck. The photo I saw of her while she was in the rescue shelter showed a shut-down little soul who seemed to have given up – she was too sad and frail to even hold up her head without help. Her paws had healed by the time she came to us, but she was very emaciated.

Saffy looked anxious about meeting Skye and Charlie when we brought them out into the road to go to the field, so we went for a long group walk around the field and through the woods. Charlie barked a lot at first, but I reminded him to focus on me each time he got over-excited, and he soon settled down and walked quietly. We gradually allowed them to move closer, making sure there was enough distance to keep them all feeling safe. Saffy showed her teeth on a couple of occasions, but, by the time we left the woods, they were all strolling together peacefully.

Off-lead introductions in the garden went well, though Saffy looked a bit stunned when Skye jumped right over Charlie – one of their favourite games at that time was playing leap-frog together. Although Charlie was rather nervous of her for a while when we brought them all indoors, he showed no aggression towards her, and it didn't take long for him to relax, go on his bed, and fall asleep.

I fed them all separately to ensure no conflict arose over food, as this is common when dogs have been starved. Saffy snaffled everything down at top speed. After dinner she did some counter-surfing, which I quickly stepped forward to block. She then removed the lid from the large bin that contained the dried dog food, and helped herself to dessert, so I replaced the lid, pushing it tightly closed, and made sure the larder door was properly shut in case she found her way in there, too. After an exploration of the kitchen, looking for any tidbits I may have left lying around, she trotted into the living room and hopped on the sofa. Saffy had made herself at home, and we all quickly fell in love with her.

Charlie and Saffy became close friends, and spent most of their time playing together. It was beautiful to watch them chasing each other, standing on their hind legs to wrestle, and rolling around on the lawn. Skye was very sweet with Saffy, and took on the role of protective older brother, watching over both of the youngsters and intervening if he thought play was becoming a little too exuberant, as Saffy played rough, and Charlie's neck – sore from the staphylococcus infection and demodex that he'd arrived with – was still healing.

Toilet-training is something I have to teach most of the dogs who stay with us, and it's always worked well within three days: Charlie 'got it' instantly when he arrived. Saffy had four accidents during the evening and night in the first three days, and then figured out that the garden,

not indoors, was the place to go. I take new dogs outside after meals, drinks, naps, and when I see signs such as sniffing, circling or restlessness. They get lots of praise and a small treat when they toilet in the desired place, and this positive reinforcement means that they always learn fast.

As the weather was mild and sunny, the door to the garden was open a lot of the time so that the dogs could play out whenever they liked. However, that was creating confusion for Saffy, because there was no clear definition between inside and outside, so I closed the door, then made a point of opening it and stepping through each time the dogs went near it. Saffy immediately 'got it' and there were no more accidents.

Saffy mugged me for my food at every opportunity in the beginning, but within three days she understood that, as nothing would be forthcoming, she may as well go and rest on her bed. She was a fast learner. Those first three days were enormous fun for Charlie!

BACK TO THE WILD

What I hadn't bargained for was the impact on Charlie that living within a group of dogs after his life in the wild would have. Five days after Saffy's arrival it was as if a switch had been flicked in Charlie's mind, and he swiftly regressed, reverting to the behaviours he'd displayed on first coming to live with us.

He slunk around on his belly if people (including me) came near; bolted in a state of panic if approached; refused to eat indoors (he would only eat at all if I left food out in the garden for him), and insisted on staying outside all of the time. At night I had to catch him, which was quite a task, and stressful for both of us, and clip on a leash to bring him into the house. It was deeply upsetting to see him so afraid of people again, especially after his rapid progress over the past months.

It took me a couple of days to figure out what was most likely going on in Charlie's mind. Charlie grew up living among a close-knit group of dogs, defending his territory against intruders who would further reduce the scant resources which kept them alive. It's likely that this is how he lost an eye. Feral dogs have a greater chance of survival when they live in groups. The alert system is more efficient, due to the numbers of dogs, hunting is more successful, and they are better equipped to drive off predators. With Saffy here, Charlie was once again part of a group of dogs. His feral nature, so recently overcome, resurfaced, and he reverted to being extremely afraid and mistrustful of people.

With Saffy here, Charlie was once again part of a group of dogs

---

Their form of play changed, too. Play between Saffy and Charlie became ultra-fast and very rough, with much boxing and rearing up that they enjoyed, though I had to step in each time Saffy's teeth made contact with Charlie's still-tender neck. This type of play left them both hyper-aroused afterwards, and Skye was firmly excluded. After a few days Skye made it obvious that he was becoming increasingly concerned about how rough the play was, and stood close by each time, watching, until he felt it was necessary to step in to calm things down. He did this by moving between them, giving a quiet 'woof,' and waiting for a few moments before walking away. His perceived role changed from companion and playmate to mediator and protector, which was stressful for him.

When it was clear that Charlie was going to need a lot of help to once again live as a domestic dog, and that I would have to start afresh at the beginning with this, I spoke to the rescue, which immediately decided that Saffy should move on. We knew she would be an easy dog to home, especially as she'd learned toilet-training and house manners with us. By then, she was even polite in the presence of food, instead of counter-surfing and mugging everyone for their meals, and she was a very sweet, immensely loveable girl. I was heartbroken at the prospect of saying goodbye, even though I'd known this would come at some point, but I agreed that it was the best decision for all three dogs.

When the person who brought Saffy to us came to collect her on a Wednesday evening, Charlie cringed and bolted, but Saffy was delighted to see him and made a huge fuss, snuggling up with him while he had a cup of tea. She left without a backward glance ... and I came indoors and cried. When I felt calmer I sat in the garden and used peripheral vision to observe Charlie to see how he would react. He stayed in the garden for a while, then came indoors of his own accord and settled on his bed.

An hour later Amelia, my dear friend and vet, called round. To my relief, Charlie went to the door with Skye to greet her, though he kept his body very low, and cringed back during the first few minutes. Amelia stayed for a drink, and we discussed the huge differences between feral, street, and domestic dogs. By the time she left, an hour later, Charlie was staying in the living room, and approaching both of us to invite strokes. By late evening he seemed almost back to the boy he was before Saffy arrived, and the next day brought still more progress in this respect.

Two days later Charlie seemed like his pre-Saffy self: the switch in his mind that went on when he had his own resident dog group

continued page 81

Charlie will only drink from a bowl in the garden.
(Courtesy Kerry James)

And above: Charlie has become the clown of the family.
(Courtesy Kerry James)

Well, shall we play?
(Courtesy Kerry James)

Charlie negotiates one of his dens.
(Courtesy Kerry James)

Charlie's eager expression.
(Courtesy Kerry James)

Skye, the wise mentor and friend.
(Courtesy Kerry James)

Hunting the ball.
(Courtesy Kerry James)

Right: Charlie's mischievous character really shines through.
(Courtesy Kerry James)

A quiet moment with Amber.
(Courtesy Kerry James)

Charlie loves to play. (Courtesy Kerry James)

Madcap Charlie. (Courtesy Kerry James)

True love. (Courtesy Kerry James)

Above: All eagerness!
(Courtesy Kerry James)

Above right: Wild boy Charlie strikes again.
(Courtesy Kerry James)

Right: Skye has been instrumental in helping Charlie enjoy his new life.
(Courtesy Kerry James)

Below: Life is such fun with Charlie. (Courtesy Kerry James)

Below right: Did someone mention treats?
(Courtesy Kerry James)

Charlie's never far from Skye.
(Courtesy Kerry James)

Two happy dogs.
(Courtesy Kerry James)

"Goodbye, and please visit us again soon!"
(Courtesy Kerry James)

flicked to 'off,' and he and Skye resumed their previous closeness, though the playful aspect to their relationship never did return. Instead, theirs became more of a father-son or mentor-student bond, and has remained that way since.

From this, I learned that, in feral dogs, the wild nature will always be there, dormant below the surface, but re-emerging if a trigger occurs. I expected it to take weeks of work to bring Charlie to the stage he'd reached before Saffy's arrival, but it took only two days, so all the careful work I had done with him during the first two months clearly helped build a firm foundation. It was a relief to see him relaxed, happy and communicating healthily again.

Saffy had the happy ending/new beginning she so richly deserved, adopted by a family with several other dogs, and has a wonderful new life, surrounded by all the love and care that she didn't receive as a puppy in Greece.

A friend in need

A few days after Saffy left, Skye became seriously ill with a gastric infection that caused fever, bloody diarrhoea, extreme weakness and debilitation. I took him straight to the vet, who put him on antibiotics and anti-inflammatories. My tall, solidly-built boy rapidly lost weight until his ribs were clearly visible, his fur grew dull and fell out, and he screamed when touched near his hindquarters. For several weeks he appeared to be close to death.

Amelia had just left work to go on maternity leave, and when the antibiotics and painkillers with anti-inflammatories had only a minimal effect, we went through over six months of trooping back and forth to the vet surgery for more antibiotics, pain relief, and a wide range of tests; arranging for additional homeopathic treatment from veterinary homeopath Paula Kunkos when the veterinary surgeons were at a loss about how to help him. It was utterly heart-breaking to witness Skye's decline, and be unable to fully relieve his suffering.

The cause of his illness was never ascertained, and no firm diagnosis could be fixed on, though one of the specialist consultants told us it was likely that Skye had polyarthritis, an auto-immune inflammatory joint condition that causes systemic illness. Skye gradually regained some of his strength, but the long illness took a heavy toll on his body and (aged only six at that time), he slipped from being an energetic, active, fun-loving dog to an old boy who could no longer run and play, and who preferred to be left alone to sleep most of the time.

Skye's mystery illness also left a legacy of arthritis, ongoing colitis, and a heart murmur. Over a year later, the medication regime has helped him to regain the weight he lost, and, between recurring bouts of colitis, he is happy and comfortable, though far less active than previously.

Charlie seemed to know that the days of racing around with Skye were over, apart from an occasional light-hearted moment in the garden on one of Skye's more energetic days. It's bittersweet to see his expression of joy when Skye is able to engage in a brief game of chase, and he greets his friends in the field with sheer delight at the prospect of a good game that goes on until they're tired; bright-eyed and panting.

ELeven NeGOTIATING SETBACKS

At midnight on a hot summer night, I glance up from the book I'm reading. Skye and Charlie are asleep. Skye is curled tightly so that he can fit into the armchair beneath the front window, and Charlie is resting in his most endearing position, with his muzzle on his front paws. He senses that I'm looking at him and raises his head, meeting my eyes with a soft gaze. We share a long look, and what passes between us can only be described as love.

It was inevitable that there would be setbacks. Charlie's new life posed so many challenges for him that the heady periods of sometimes steady, sometimes rapid progress, were often followed by days – and then several months – of tough terrain for all of us to negotiate. During the first seven months, Charlie's regressions to skittering around, flattening to the ground, and bolting were all due to fear. Obvious triggers were household noises, especially the vacuum cleaner, washing machine, and the sound of unexpected voices or doors closing – these all pose a problem for many dogs, so were no surprise to us. People approaching him too fast or head-on sent him into a state of panic that took several days to recover from. And, of course, his regression to his former wild, extremely fearful state when I fostered Saffy was deeply saddening, though, fortunately, he soon overcame this and life returned to normal.

A very negative experience with another young Whippet mix dog who disliked Charlie intensely sent him spiralling back down, but, again, after a few days he recovered and life carried on as normal. Charlie had adjusted better than anyone expected, and was very popular in the village by then.

Seven months after coming to us, Charlie showed signs that his confidence had really blossomed. He strutted along, head up and tail so high it curled over his back like a Husky's, and he was less fearful of new experiences. We were delighted until the repercussions of this hit home. Charlie became a bully, and this rough patch lasted for over six months. The sudden shift into aggressive behaviour towards all of us, including Skye, was most likely triggered by Charlie sensing that I was going through a state of emotional upheaval. Always sensitive to the moods of those around him, his perception of me as weakened prompted him to react aggressively, and his behaviour became very threatening. During these months Charlie was a danger to those around him and needed very careful monitoring.

The match that ignited a stressful time for me, and a period of unpredictable aggression in Charlie, was my determination to raise public awareness of the worrying illegal importation of unvaccinated dogs, as almost 8000 had been brought into the UK from overseas for adoption in 2013, some of them on falsified pet passports.

During that year I had founded the Dog

Welfare Alliance, which brings members of the public together with professionals who work with dogs and subscribe to positive methods. When possible, the Alliance contributes funds to rescues around the world who need urgent help. An experience with a Romanian rescue that September, whose cavalier attitude towards rabies vaccines made it clear that unvaccinated dogs were being imported into Europe and the UK on forged pet passports, set warning bells ringing. This flouting of the law was dangerous as well as illegal, and the possibility of dogs arriving who were not vaccinated against rabies posed a serious health risk to people and all animals. Discussions with animal welfare officers, the veterinary director of Dogs Trust charity, and numerous rescue volunteers who were dealing with the backlash of this resulted in my speaking out strongly about the need for overseas vets and rescues to ensure that dogs are properly vaccinated, and for tighter border controls when it came to checking pet passports.

During my four-month intensive investigation, word spread through social networking, particularly Facebook, that I was writing an article about illegal importation for *Dogs Today* magazine. I had already written an article for the Dog Welfare Alliance website titled 'Points to Consider Regarding Importing Rescue Dogs,' which outlined my main concerns: the number of rescues springing up, run by individuals who were hard to identify, yet were using social networking to ask for donations; the issues with falsified passports for unvaccinated dogs; the health issues of many of the dogs, which had turned out to be prohibitively expensive for some adopters who were unaware of these until their dogs arrived, and the behaviour issues with some of the free-ranging dogs due to lack of socialisation.

In this article, and the one I wrote for *Dogs Today* magazine, I made it clear that I was not opposed to importing dogs, and had, in fact, adopted a Romanian feral dog, but that stringent application of the laws was vital; that imported dogs should come with rescue backup in case anything went awry, and that potential adopters needed to be aware of the responsibility and commitment that would be required when taking on such an animal.

A furore ensued, and I received a lot of hate mail and abusive phone calls

---

A furore ensued, and I received a lot of hate mail and abusive phone calls from importers. Some rescue organisations, who were dealing with the emotionally-damaged, hard-to-rehome dogs surrendered to them by adopters who could no longer cope, staunchly defended my stance, whilst other rescues, who imported large numbers of unsocialized dogs, led an outcry against my article. It was a difficult time, but I felt strongly that this information should be made public, and I was glad that Beverley Cuddy, Editor of *Dogs Today* magazine, gave me the opportunity to bring the issue into the open. However, the stress this created had a significant impact on Charlie.

JEKYLL AND HYDE

Charlie would be sweet and affectionate one moment; then, the next, would rise from his bed and pad over to stand very close to me, uttering deep, low growls, with his commissure tightly pursed, his body rigid, and his eye hard and vacant. During these episodes, the Charlie we knew and loved was temporarily absent. This behaviour occurred daily numerous times for

Charlie

over two months, and then frequently for several more months, without any evident trigger. Even though I've worked successfully with many domestic dogs with extreme fear and aggression issues, Charlie had me and my professional colleagues flummoxed.

Friends stopped visiting because his changes of mood were so sudden and unpredictable. Visitors were targeted, and I could understand why they preferred not to take the risk. It was painful, a dark night of the soul, and I talked in depth with several friends and colleagues who are experts in wolf and dog behaviour, searching for reasons for Charlie's Jekyll and Hyde-type shifts.

One hypothesis was that his current life was so alien to his background that there was a conflict between his wild self and his 'domesticated' self; that Charlie's nature had been divided in two as cleanly as an axe splits a log. A thorough veterinary check and array of tests ruled out any clear-cut health issues that could cause these extreme shifts in mood, and I explored every avenue I could think of, seeking

---

My wild boy had strongly bonded with us

---

ways by which to heal this rift.

After seven months and the formation of a profound bond, it was necessary to go back to the beginning and start afresh with Charlie. The dog-like aspect of his nature was easy to understand, as was his need for a sense of connection, for love and physical contact, and for the acknowledgement of who and what he is. Yet Charlie's wild self, which he lived and fully expressed until he was taken into rescue, was another landscape altogether. Looking into Charlie's eye I could see a wise and ancient soul

who, at that point, regularly struggled against the comparatively luxurious constraints of captivity.

My wild boy had strongly bonded with us – of that I have no doubt. His delighted greetings when he woke in the morning and when I returned home after a short trip out; his affectionate interactions; his playful moods and tongue-lolling squirmings at the prospect of a stroke, walk or meal, all indicated how deeply his bond with us was. The shift in his psyche occurred without warning: our usual connection shimmered briefly, then was gone, as he stood four-square, furiously intent on doing harm. The terrified soul of the early weeks, and the soft, affectionate dog who emerged from that sad chrysalis, thanks to much careful nurturing, was temporarily subsumed. Instead, the Charlie who faced me down was a soul who had, once again, heard the wild call, and apparently bitterly resented his captivity.

Speaking to him elicited no response, because the Charlie who knew his name was no longer there. My only option then was to break the spell, if he was far enough away, by dropping a pen on the wood floor to startle him and snap him out of it – an aversive action that went very much against the grain for me. When he was too close beside me for this, I turned away, made myself as small as possible, and twice, when he tensed to spring, had no alternative but to calmly leave the room.

Each time, after a few minutes, his mood changed again: the strange animal with Charlie's features blinked ... and Charlie returned, to walk away and lie on his bed, tightly curled with his back turned to the world. Later, after resting, he would come to beg for a stroke and loving words, and the strange cycle would begin all over again.

Several people who had adopted free-ranging dogs contacted me at this time, all with the same story; all of them hurt, confused, upset, and at a loss as to how to deal with the sudden, inexplicable onset of aggression that they were experiencing after their dogs had been with them for several months. A dog who was rescued alongside Charlie had bitten his carers several times; others had escaped and vanished. Some carers, aware that their dogs could not be rehomed, were fearful that their companions would end up on a one-way trip to the vet for a lethal injection. I was fortunate that I could read the signs early enough, and instantly act on them, to avoid being bitten by Charlie. The common thread through all of these conversations was deep, painful sadness. We had all shed tears. We loved our dogs, and they loved us – when they weren't overcome by an urge to attack us. Most of us had no inkling of their background before they arrived in our homes. We all felt we had played a part in inadvertently imprisoning them, and that our dogs were paying a very high price for the comfort, warmth, regular meals, shelter, and affection we wanted to give them.

There were times when I questioned both myself and the wisdom of rescues that import free-ranging, unsocialized dogs into an environment they may be unable to fully adjust to. I have never regretted adopting Charlie, but there were times during this period when I knew that, had I been aware of his background, I would have more carefully considered what the impact would be on all our lives. I worried about Skye's now-frail health, and felt guilty that this gentle dog, who had given so much of himself to us and to all the other dogs who lived with us, was having a hard time because of Charlie's unpredictable behaviour.

Enter Team Charlie

By New Year 2014, Charlie's aggression had escalated to such a level that my daughter rarely entered the living room, and she worried about going out and leaving me alone with Charlie in case she returned to find me badly hurt. A number of people, dog professionals included, strongly advised me to have him euthanized for our own safety. I had to accept that, if nothing else worked, this may have to be considered, but even the thought of it was unbearable.

I was sure that there must be some way through, and I put together Team Charlie to explore every available option. This consisted of a group of experts, which included my two regular vets, Amelia and Liz, Tellington TTouch expert Sarah Fisher, wolf and dog expert Toni Shelbourne, homeopathic vet Paula Kunkos, holistic vet Sarah Drawbridge, and my close friend and behaviour colleague Theo Stewart. Additional emotional support came from my friend Lisa Dickinson and my cousin, Sue Beech. We all discussed the situation in depth, and a plan of action was drawn up and implemented.

My vet put Charlie on a course of Zylkene, a nutraceutical that can have a calming effect. I added Bach Flower remedies to his food and water, and used a Pet Remedy plug-in in the living room, and Pet Remedy spray on his bed, the furniture and rugs. Sarah Fisher suggested I change Charlie's diet to fish and potato for a couple of months, in case he was reacting to any ingredients in his food. She also introduced me to Sarah Drawbridge, who visited and taught me some basic Tellington TTouch movements, as well as using Zoopharmacognosy, a system that involves letting Charlie self-select several plants and oils from a multitude that were set out around the room in small jars and bottles. Sarah impregnated a cloth with Charlie's choices, and I kept this under his bed so that he would inhale the scents. Paula Kunkos treated him with

Charlie

homeopathic remedies.

Gradually, Charlie began to calm down. The aggressive episodes carried on for a few more months, but became less intense, and the time between them grew longer: at first they occurred every few days, then weeks, then a month, and over the past months there have been no serious recurrences. Any complacency would be foolish, but life has been much smoother for all of us recently, and I'm immensely grateful to

---

Charlie's nature is, and always will be, wild at heart

---

everyone in Team Charlie for all their help.

Experiencing this darker side of Charlie, and hearing about similar experiences from others who had adopted free-ranging dogs, led me to consider how dogs with a similar background could be better helped in their countries of origin. The Dog Welfare Alliance now supports several neuter and release programmes overseas, especially in Romania and Bulgaria, as this will ultimately reduce the widespread populations of dogs who are either starving or are being killed in the streets. Dogs who would not be able to adjust to a home environment can then live the lives they are used to and are well adapted to. This will put less pressure on the rescues to find spaces for dogs who are considered unhomeable, and will enable them to home those dogs whom they have assessed as adoptable.

I am fully aware that there may be some rocky patches to traverse in the future, and that any setbacks are simply strong reminders that Charlie's nature is, and always will be, wild at heart. Our love for him and acceptance of him must be unconditional, while at the same time we take steps to ensure that he remains safe to be around.

Over the past few months we have seen a major shift in Charlie. He has become more like a domestic dog, though it would be unwise to think that his wild self has permanently gone deep underground. His gaze is usually 'soft' and gentle, rather than hard and challenging. He has become increasingly more affectionate and playful towards us, and towards visitors. Even his face has changed and softened, with the previous hard edges of tension erased.

Together, we faced the worst, and made it through. The bond between us has grown even stronger.

TWELVE we can all be Heroes

Charlie has dug two dens in the middle of the lawn. One is a cave, wide enough for him to curl up and lie down in; the other is a narrow tunnel with a triangular outer edge that he likes to drop squeaky balls into before disappearing inside to retrieve them. All that can be seen is his rear end and gently wagging tail as he scrabbles in the mud to extend his tunnel further.

In Romania, Charlie would have made dens in which to take shelter during rough weather, and keep cool in summer. Skye has learned to negotiate Charlie's excavations after a couple of incidents when he ignominiously slid into them while playing tag with Charlie. After numerous games that involved me laboriously carting earth across the garden to fill in the dens, then watching, laughing, as Charlie immediately dug them out again, I've given up on any attempt at gardening pride, and simply mow around the edges. I just thank my lucky stars that the dens are angled away from the foundations of the house ...

Charlie knew how to keep himself safe from harm in the wild. He made dens to protect himself from the elements. His hunting skills provided him with food, and, considering that, although he was thin he was certainly not emaciated, as other rescue dogs have been when they arrived, he was clearly good at fending for himself. During the first two months here he stalked insects, slapping his paw on them before devouring them, and the problem we'd had with hordes of woodlice sneaking in and taking over the house was soon vanquished. He had less luck with birds, though that was more to do with me shooing them away to safety before he could pounce. Now that he's well fed, he shows less interest in catching his own meals, though it's likely that he would still be able to survive on his own again if he had to.

His territorial instinct is very strong

His territorial instinct is very strong, and he would have staked out his own area in the wild; marking the boundaries, as he does here, and driving off intruders. Some nasty scars on his left hind leg – and, of course, the loss of his left eye – tell tales of battles fought in the past. He functions so well with only one eye that we tend to forget, and think of him as being fully-sighted – especially as the surgeon who operated on him in Romania was able to leave his eyebrow and the surrounding nerves intact, so both sides of his face remain mobile. During Charlie's roughest few months here, when he suddenly began to exhibit frequent aggression, we wondered whether, perhaps, he had sustained some kind of brain damage from infection or surgery, but tests showed that it was more likely to be a psychological rather than a physical issue.

Charlie

Domestic life was so alien to Charlie that the skills he had developed for coping in the wild were of little use when he was continually faced by situations way beyond his understanding. His freeze, flight or fight responses were all that he had to fall back on, and this new world he'd been thrust into was such a confusing place that he reacted to absolutely everything. It was my responsibility to help Charlie feel safe; to be his champion and protector.

The modern world is a dangerous place for dogs – even for those who were born into it, and who learn to accept the bombardment of the senses that they're continually subjected to. They need to be protected from traffic, household objects and chemicals that could harm them, and negative attention from people and other dogs, among many other things, and it's our job to teach them how to live safely and harmoniously alongside us in a society that's now governed by technology. It's also important that we're aware of the impact that life within our homes has on them.

superior senses

The dog's sense of smell is at least fifty times more powerful than ours, and he sees more widely than we do with our 180 degree vision (a dog's visual range is 250 to 270 degrees). Images are also translated faster. Still snapshots are transcribed by the brain into moving images; this is called the flicker fusion rate. Our human flicker fusion rate is sixty cycles per minute, whereas that of dogs is between seventy and eighty cycles per minute. They can spot the slightest movement before we're even aware of it, and can map the trajectory of a moving object far more accurately than can we. Their hearing is far superior to ours, reaching the ultrasonic range of frequencies of up to forty five kilohertz compared to our highest auditory range of twenty kilohertz. In short, Charlie was utterly overwhelmed by the constant assault on his senses of a multitude of unfamiliar smells, sights and sounds – all at high volume.

In the wild, Charlie's survival depended primarily upon his ability to locate food, avoid predators, and find shelter. The scents, sights and sounds he picked up were those of nature. His brain processed the world around him, categorising wind, rain, snow and ice, the terrain beneath his paws, the sound, scent and glimpse of prey, bird cries, the rustling of trees and grass, and the movements of other dogs in and beyond his social group. His life was hard, in the sense of enduring bad weather and times of scarcity when little food was available, but it was the life he knew and was perfectly adapted to.

In contrast, the buzzing of electric wires in the walls, way beyond my range of hearing, was terrifying for a dog who had never lived indoors. The plink-plink sound of raindrops hitting the shelf inside the fireplace, and the glass in the windows must have been an assault on his ears. The smells of cooking, of human scents, of perfume, of olfactory traces left on the furniture by previous guests, human and canine, of urine marking from passing dogs on the outside of the gate and hedge, and many other scents that I was no doubt unaware of, were deeply confusing and worrying for him.

Compounding his anxiety even further were the day-to-day occurrences that Skye and other dogs accepted as normal. Doors being opened and closed; the clink of food and water bowls; the chinking sound of his identification tag as it rang out against the edges of these; the unfamiliar foods which smelled like nothing he had encountered before; the sound of human voices, pages being turned in a book, music, children running past, cars starting up, lights being switched on and off; the vacuum cleaner, washing machine and tumble dryer.

In addition to these was the terror Charlie felt when unfamiliar people came to our home or approached us outside, and the shock of seeing many other unknown dogs in an area where he had yet to figure out the boundaries of his new territory.

we can Be Heroes

Our dogs rely on us to take care of them, ensure their safety, monitor their health and meet all their needs. Their lives are ruled by us. On the whole, we decide where they will sleep, when, where and what they will eat, who they socialise with, when and where they can play, and where they exercise. The quality of their lives is dependent on the decisions we make on their behalf. It's no wonder that dogs who lead very restricted lives, where they are left alone for long periods and expected to interact only when their carers have free time, often end up with behaviour issues. We have a duty of care to our canine companions, and this includes ensuring we take steps to understand their natures, and allow them opportunities to express these as fully as possible: to let them be dogs, while supporting and standing up for them in situations where they need our help and protection.

... any approach was a rude intrusion on his personal space

Dogs need us to be their heroes: to be courageous on their behalf; to speak out for them because they are unable to speak out for themselves, and to take immediate action to protect them when their safety or wellbeing is compromised. Even more than any other dog I

had lived and worked with, Charlie needed me to be his champion, and I stepped up to the mark constantly during his first year in my care.

I'm a naturally friendly, considerate sort of person: I like people, and I strongly dislike being in a position where I have to be rude, but there were times when bluntness was necessary in order to protect Charlie from attention that worried and frightened him. Charlie is a beautiful dog; his appearance is such that, when we go out for walks, it attracts people who want to stroke him. He's the size of a small Labrador, so isn't intimidatingly large, and his coat shimmers as if sprinkled with glitter, reflecting light that sparks out around him on sunny days. These factors, combined with his unusual reddish colouring and markings, his soft floppy ears, and his single pale amber eye, prove an irresistible draw for a lot of people.

To Charlie, however, any approach was a rude intrusion on his personal space: as unpleasant and unnerving for him as it would be for me if a stranger suddenly walked over and grabbed me by the shoulders. Charlie found such interactions terrifying, and would either launch into full voice, complete with 'back off' signals such as growling, lunging and hackling, or cringe back, trying to hide behind me. Essentially, he would exhibit every behaviour in his repertoire that might possibly give the desired effect of keeping a safe space between himself and whatever he viewed as a threat.

I became adept at stepping quickly in-between Charlie and the offending individual, shielding Charlie with my body, whilst calmly but firmly explaining that he was a very scared feral dog with no previous experience of people. There were even occasions when I had to hold up my hand, palm out in the traffic cop 'Stop!' position when someone bore down on us at speed with

Charlie

a determined 'I adore dogs' gleam in their eyes.

On the few occasions where people kept on coming, despite this, asserting that all dogs love them, I had to move between them and Charlie and tell them that he may lunge, or even bite, if they came too close. It was easy to assess the limit of Charlie's 'safe zone,' the area of space he needed around him in order to feel comfortable. As with my friends in the village, I asked strangers to remain outside Charlie's safe zone, and most people were willing to comply, and step back a little to ask me about him. If anyone persisted in intruding, I simply walked Charlie away. With those who were considerate, stopping to chat led to some interesting discussions, and the distance between us meant that Charlie felt safe enough to begin to relax and accept their presence. I told people that they were helping a great deal by being there without putting him under pressure, which always elicited a positive response, and often the discussion would end up being about their dogs, past and present. Ultimately, they went on their way, having taught Charlie that they were no threat to him, and often feeling rather good about themselves for helping a very scared dog to cope.

Learning, in this way, that I would protect him helped Charlie develop more confidence around new people, and those who had met him several times at a safe distance were immensely flattered when he gathered the courage to step forward and greet them. When this happened, I asked them to move slowly and keep their hands where he could see them, avoiding his head area if he invited a stroke. It was heart-warming when he recognised friends in the distance, and became eager to say hello.

As Charlie grew more confident, a new protective instinct emerged, and his territorial boundary included me when strangers approached us. I needed to remind him that it was my job to protect him, not vice versa, and the strategy I'd used when he was fearful of unknown people again came into play. Keeping a safe distance and rewarding Charlie for focusing on me, rather than the other person, worked well, as did employing body language that showed him I wasn't concerned about their proximity. This involved keeping his leash loose, and stroking my fingers down it as though I was stroking him, because our emotions are transmitted down the leash to our dogs. I turned my body slightly sideways and made sure that it was fluid, with no areas of tension, to show Charlie that I was comfortable in this situation, so he could relax, too.

Although they have otherwise been the best of friends, there were times when Skye needed me to protect him from Charlie. When Charlie went through his aggressive period he targeted Skye on a number of occasions – mostly when Skye was lying down, resting. Skye, being a gentle soul, stayed perfectly still when Charlie stood over him, growling; turned his head away to signal that he had no interest in conflict, and moved quietly out of the room to go upstairs and rest on my bed each time I diverted Charlie's attention onto myself. Fortunately, this phase passed, and they're both very close friends again.

Defining Territory

Nowadays, Charlie's territory has narrowed to our garden, instead of stretching as far as his eye can see. He noisily charges around when unknown or unfriendly dogs go past, howls or makes a strange chimp-like noise if he hears his dog friends nearby – his way of calling them to him to play – and his only source of regular daily irritation became the chipmunks next door.

In springtime, Cath and Phil erected a new outdoor home for their chipmunks. It's tall, much higher than our boundary fence, and it drove Charlie to distraction to see these sweet, fluffy little souls running up and down the mesh on the side of the cage that faces our garden. It was clear that he viewed them as prey, and engaged in futile attempts to stalk them, barking in frustration that they were always out of reach. This is one hunting endeavour that he would never have an opportunity to engage in. It took over three months for Charlie to accept that the chipmunks were there to stay, and that shouting and trying to reach them was a fruitless exercise. All became quiet in the garden once more, with Charlie redirecting his hunting instincts at the rat holes in the woods, digging his way in, though, fortunately, the rats were long gone.

Charlie has learned that he is safe, that he can relax and enjoy company, because it's his choice whether to interact with people or to move away, and that the choice is respected. This release from any pressure has helped facilitate the change to the friendly, happy, well-adjusted dog he has become.

VISIT HUBBLE and HATTIE ON THE WEB: WWW.HUBBLEandHATTIE.COM
WWW.HUBBLEandHATTIE.BLOGSPOT.CO.UK
· DETAILS OF ALL BOOKS · SPECIAL OFFERS · NEWSLETTER · NEW BOOK NEWS

91

Thirteen
The Healing Power of Laughter

For months, Charlie has been terrified of the lawnmower. Even the sound of it starting up sends him into a quaking, floor-hugging mess, and each time it takes him several days to recover. This is unfortunate because the grass grows so fast here that the lawns have to be mowed at least once-weekly to avoid having a jungle instead of a garden.

Then one day Charlie ventures outside with Amber and Skye whilst I'm in mid-mow.

Our garden slopes steeply down from the house, which makes grass maintenance very warm work, and excellent for the rediscovery of muscles left to languish all winter. It also means that toys roll downhill unless they've already been deposited in one of Charlie's dens. On this day, Charlie suddenly sees the potential for a new game. The lawn is strewn with dog balls, toys, and a few stag bar chews that I nudge out of the way with my foot as I go along. Charlie dives on the nearest ball, races over to me, and drops it right in front of the mower. I stop and throw the ball to the other side of the garden. Charlie's eye lights up and the game begins.

Every object is carried across, one at a time, and deposited right where I'm aiming the mower. I shift them aside with my foot, concerned that Charlie will come too close, and they roll down the slope. He races off to find the next toy and drops it in front of me. It takes an hour longer than usual to cut the grass, because I have to keep stopping the mower, and I'm weak from

laughing at Charlie's expression of mischievous glee each time he brings something else to me.

Laughter is such good medicine. It defuses tension, and blows away the cares and worries that can so easily wrinkle the fabric of our days. It puts life's ups and downs in proportion. And even the darker moments have their funny side if we step back and look at them from the outside. I've found it's the best way to negotiate the challenging times, and we've enjoyed a great many belly-laughs at Charlie's peculiarities and idiosyncrasies.

A very cheeky, puppyish side of Charlie's nature emerged

A very cheeky, puppyish side of Charlie's nature emerged once he overcame many of his fears. Skye, who *was* a madcap puppy, and is now a wise, steady adult, only occasionally has the energy for fun since his long illness left him very frail, so Charlie became the canine clown of the family. Skye joins in when he feels able to, but mostly he watches from his resting place, his expression very similar to that of an elderly person indulgently overseeing a mischievous child. Sometimes I get the impression that Skye is inwardly shaking his head and muttering "Here he goes again!" Yet he still takes the time to engage in tutoring when he thinks Charlie needs help.

Charlie developed a penchant for helping himself to Amber's newly-washed socks. He silently stalked her when she removed her clothes from the tumble dryer, peeking around the corner like a cartoon character, waiting for her to drop one – which she invariably did – on her way up the stairs. Instantly, he would pounce and run off, tail high, hugely proud of himself, before finding a quiet spot and tearing the sock to shreds. Amber soon had a collection of sock singletons, and my supply was depleted when she 'borrowed' matching pairs from me, which also fell victim to Charlie's sock fetish. A new house rule was established: watch your socks, and don't let Charlie near them!

---

*Watch your socks, and don't let
Charlie near them!*

The strip curtains I put up outside the back door to keep out flies provided another opportunity for Skye to demonstrate his teaching skills. Charlie was terrified of these, and refused to go outside until I held them open for him. Of course, he also waited for me to create a clear entrance when he wanted to come back inside. After a couple of days – given that Charlie likes to trot in and out frequently – I found I was up and down from my desk every few minutes, feeling like a human Jack-in-the-box. Apart from anything else, the curtains weren't fulfilling their purpose, and the constant distractions weren't conducive to finishing the novel I was writing. Which was when Skye stepped in and decided it was time Charlie figured it out for himself.

Skye casually strolled through the curtain, then stood on the patio and looked back over his shoulder at Charlie, who hovered nervously in the hallway. Skye turned and strolled back indoors, glancing sideways at Charlie as he walked into the kitchen. Charlie joined us in the kitchen, and Skye gave me a look that could so easily have been interpreted as "Oh, good grief!" He walked back through the curtain into the garden again; then returned.

This went on for a while. Suddenly, Charlie made a dash for it, through the curtain at speed, and ran around the garden doing a lap of victory. Coming back indoors was too much of a challenge, however, despite Skye's patient tutoring, so I held back the curtain and he raced through, skittering in as if he thought it would attack him.

It took three days for Charlie, encouraged by Skye, to understand that he could come and go easily, with no reason to fear that the silvery strands would attack him. After that, there was no stopping him.

Squeaky toys posed another challenge for Charlie. The first time Skye repeatedly held and released a squeaky ball, Charlie bolted, after a while, crawling back in on his belly, anxiously watching Skye from a safe distance. Skye rolled the ball across to him and waited, ears pricked with anticipation. Charlie pounced, then leaped back as the ball squeaked under his paws. Skye cocked his head as if to say "Your turn!" Charlie pounced again, imitating the sound of the ball with high-pitched yelping. He batted it with his paw and it rolled towards Skye, who grabbed it, squeaked it, then batted it back to Charlie. After several more rounds Charlie figured out that this was a fun game, and threw himself wholeheartedly into it, an expression of intense focus writ all over his face, while Skye trotted off to find another ball, leaving Charlie to carry his around, squeaking it continuously.

SELF-EXPRESSION

Charlie's vocal range is broad. He howls at the

sound of the printer, for example, and for a long time would howl every time the phone rang. When he's excited, he makes a strange yipping noise that sounds more like a chimpanzee than a dog, especially when we set off on a walk. He purrs when he's snuggled up beside me on the sofa with his head in my lap. His bark sounds like a rooster crowing, and it carries a long way. His growl is low and resonant: a clear warning to back off that has immediate effect. And when I return home after a short absence, he whines as he rubs himself against every part of me that he can reach, his tail following the curve of his body to ensure I'm well and truly infused with his scent. Charlie constantly makes me smile, and his facial expressions are often so comically eloquent that he has us laughing out loud.

The word 'treats' has Charlie pricking his ears and racing through at top speed to wait by the larder, his face a picture of anticipation as he dances on the spot. The rustle of the cheese wrapper has him in paroxysms of sheer delight, and his open-mouthed smile and hopeful expression is hard to resist. It's easy to read him, as the shape of his face changes according to how he's feeling. His face narrows, the skin tightening against his skull, and his eyebrows furrow when he feels tense or worried, and widen when he's relaxed. His eye, with its distinctive eyeliner drawn right across to blend with the darker fur around his ear, shifts from almond-shaped to round, depending on his frame of mind. At times his face resembles that of a Labrador; other times it's more like a Bull Terrier. Sometimes it appears leonine, and often, especially when his prey drive is activated, he looks distinctly wolf-like.

Watching Charlie practice his hunting skills is as fascinating as it is entertaining. Spotting a ball on the lawn he stalks it, and, once close, pounces, repeatedly hammering down his front legs, his entire body bouncing, ears angled forward; a totally focused expression on his face. In those moments nothing else exists in his world. He's utterly absorbed, alight with intensity and sheer wild joy as he stomps the ground, grabs the ball with his mouth, shakes his head, drops it, and begins hammering the ground again.

I spoke to my wolf expert friend, Isla Fishburn, about this, as it occurred to me that, whilst living wild, Charlie would most likely have used this hammering motion to drum out rodents from underground, and then pounce to devour them. Isla was excited about this evidence of the wild-domestic link, and told me that one of her wolfdogs, Tunkasila, catches mice in this way when they're out on country roads. Tunky leaps into the air, pounces, and smashes both of her front paws on the ground. This shocks, and most likely kills, the mouse, and then she eats it. As Charlie lived within a small group consisting, possibly, of just himself and Lennie, the dog who was his constant companion each time his Romanian rescuer, Denisa, observed them, small prey would have been both easier and safer to catch than taking the risk of being injured by large animals.

THE JOY OF FRIENDSHIP

The spontaneity of non-human animals can teach us so much about the joy of living in and experiencing the moment, without thinking about yesterday or tomorrow. I've felt privileged to share my life with so many friends of different species.

The many dogs I've loved and lived with began with Bobby, a Border Collie adopted from the local shelter when I was seven years old, who strolled through the woods with me, just the two of us, and lay beside me under the trees while I wrote poetry and stories. The echoes from that time reverberated when I watched my young children run, roll and tumble with our dog, Carnie, in our garden in Ireland. The chameleon, whose fly-catching abilities awed me in Malta, and the monkey we took care of while living in Singapore, opened up new worlds in my mind. Swimming alone with a wild dolphin in the cold, cold waters off the coast of Ireland overwhelmed me with emotion when I gazed into his extraordinarily wise eyes, before he dived and came up beneath me to raise me on his back. The warmth of companionship was experienced with Bhakti, my cat, when he curled around my neck like a scarf while I painted the illustrations for a book. The indescribable thrill of flying a hawk, feeling the sudden weight as he returned to my wrist, and the delight of standing, arms outstretched, while a flock of doves flew in to land on my arms and shoulders in the park. Being befriended by a tortoise who followed my voice and came to lie on my foot – each of these experiences made my heart sing.

And now it is Charlie and Skye who keep that heart-song alive, reverberating in that secret place untouched by the cares and vagaries of the roller-coaster ride that is life. Although Skye is fragile in body now, his spirit is as strong as ever. Charlie understands that their days of rough-and-tumble have passed, and he finds other ways to express his pleasure in Skye's friendship: muzzle-fencing (a game in which both dogs nudge and bat their muzzles against each other: a bit like a sword fight with noses!) in the garden; checking that Skye is not too far behind in the field, and running through to fetch Skye when he senses the irresistible presence of cheese so that they can stand side-by-side, and share.

These small beauties are the stuff of life that have the power to send my spirits soaring. The sight of a young feral dog taking care of the friend who has taken such good care of him as he high-steps through his new life, learning daily to embrace it even more fully with open paws, is a vision of pure joy.

Fourteen FINDING STRENGTH

Five months after Charlie's arrival, Amber sits in the armchair beside my desk and we chat while I take a few minutes' break from work. Skye comes over to request an ear rub, and Charlie follows close behind, nudging Skye with the side of his head to get him to move over and make room. He's becoming quite cheeky lately, butting in to make sure that he doesn't miss out on any of the affection that's being shared around.

Ear rubs and mini massages duly given and gratefully received, Charlie hesitates in front of Amber before furtively clambering up onto her lap. It's a tight squeeze. He hangs his head low and avoids looking at her, as if trying not to draw attention to himself. Our eyes meet, mirroring our mutual astonishment. Charlie has joined me on the sofa a few times, employing that same hang-dog expression each time, as if he's worried that he may not be welcome, but this is the first time he's made this move towards Amber. It's yet another breakthrough.

In learning to circumnavigate a totally alien environment to the one in which he grew up, and considering the multitude of fears he was beset with during the first year, Charlie has proven to be one of the strongest, most resilient creatures I've had the honour to meet. The sheer magnitude of the challenges he was faced with would be overwhelming for many of us, but he has overcome them in a manner that would most likely be described as heroic if he was human.

Of course, he's had a great deal of help from us; from Skye, and also from Team Charlie during his roughest patch, but, ultimately, it was Charlie who had to make those inner shifts that brought him through from a continual state of terror to tremendous achievement.

Not all unsocialized feral dogs learn to adapt and cope as well as Charlie has

---

Although many of the street dogs who are placed in homes adjust quite quickly – especially those already accustomed to the presence of people – not all unsocialized feral dogs learn to adapt and cope as well as Charlie has. I've heard from carers whose imported dogs still, after several years in a home, avoid human contact, and bolt at the sight of a leash. Their lives are permanently restricted to only certain areas of the house and garden.

This sad situation has prompted me to consider how we define 'rescue.' A life of anxious, self-imposed isolation in a home, however comfortable the bed or nourishing the food, hardly constitutes happiness for a free-ranging, social being. Welfare, whether of human or non-human animals, encompasses psychological and emotional, as well as physical, needs.

The question of what it means to place feral dogs in homes hadn't really occurred to me before Charlie came to live with us. The feral dogs in the areas of Singapore and Malaysia where I lived as a teenager were happy in their free-roaming lives, and when my sister and I begged our parents to allow us to take in the feral dog we'd befriended, their response was that it wouldn't be fair to do so. Now, having seen the tough challenges this posed for Charlie, and in view of other dogs from similar backgrounds being imported for homing, it's very clear that potential adopters need to be aware of what this actually entails for the dogs and their new carers. Carers must possess certain qualities if the dogs are to be successfully habituated to the new environment: an understanding of dog behaviour, patience, compassion, a willingness to make possibly drastic changes in lifestyle, and the strength and staying power to keep going when the road gets rocky.

A feral dog has to be strong to survive in the wild. Starvation, disease, predators (including humans), and conflicts between different social groups of dogs pick off many individuals before they reach maturity. That strength of will and determination; that extraordinary survival instinct, is just as powerfully present in the new domestic home as in the previous free-ranging environment. Charlie was very shut down when he first arrived, yet his resilience and extraordinary inner strength made it possible for him to adapt, little by little. He is a survivor.

Facing Down and Stepping Up

When Charlie is unable to remove himself from something that makes him fearful, he faces his demons head-on, and this fierce aspect of his nature is awe-inspiring to behold. His initial reaction to the television and my guitar sound amusing when described in retrospect, but at the time I admired him for challenging the one-eyed monster in the room, and making it clear he would not back down, however scared he was. Despite, and perhaps even because of, his many fears, Charlie is one of the strongest souls I have ever met.

One of the qualities that connects Charlie and I is the ability to adapt and overcome, though this occurred to me only after he had lived with us for sixteen months. My family moved home a great deal, and I went to fourteen schools altogether. We lived in England, Scotland, Malta, Malaysia, and Singapore, and later I moved to Ireland. This nomadic life was stimulating, sometimes scary, always enriching and educational, and often exciting. It encouraged the development of flexibility, a quality that has come in very useful while making the necessary changes to accommodate Charlie's needs. I've had narrow scrapes with death several times and returned from the brink, which has bestowed an appreciation for each moment of life, alongside the understanding and acceptance that our lives are finite. Like Charlie, I am a survivor, and perhaps this unconsciously linked us as kindred spirits from very different worlds.

Our inner resources are gifts which enable us to make the most of every moment, whatever that moment brings. A calm, cool head in the face of danger, the strength to meet life's challenges head-on, the patience to endure, the compassion and empathy that enable us to feel deeply for others, the determination to persevere instead of throw our hands up in surrender – all of these qualities are there inside us, just waiting to be called upon when needed.

These are the strengths of the wild self within each of us. They go hand-in-paw with spontaneity, with joy, with playfulness, with love in all its forms, with the sheer delight of breathing in the scents that drift on the air and with feasting our eyes on the sights that lift our spirits; in the

Charlie

purity of birdsong and the rallying, full-throated howl of a feral dog calling his friends to come join him.

Charlie has taught me a great deal. He has reminded me that we are all capable of far more than we realise, and that any challenge, however difficult it may seem, can be overcome if we only lighten our steps and choreograph the way through. As for what I have taught Charlie – well, my hope is that I have shown him what it is to be loved and respected by a being of a different species. He has learned to enjoy the pleasures of comfort and security, and he is able to be himself, with no conditions attached. Strength and affection grow and flourish in the fertile ground of acceptance.

VISIT HUBBLE AND HATTIE ON THE WEB: WWW.HUBBLEANDHATTIE.COM
WWW.HUBBLEANDHATTIE.BLOGSPOT.CO.UK
• DETAILS OF ALL BOOKS • SPECIAL OFFERS • NEWSLETTER • NEW BOOK NEWS

98

FIFTEEN BUILDING CONFIDENCE

There is only one street lamp in our entire village, so any night-time illumination out of doors comes from the soft glow of the moon. As his fur is now mostly a dark grey colour, Skye blends into the background when he goes out to the garden for a late-night ramble.

Nine months after Charlie's arrival, he and Skye step outside for a midnight sniff around, their ears swivelling as they decipher the calls of a nearby fox and an owl. They come back indoors together while my back is turned in the kitchen, and a couple of minutes later I step through and close the door.

As Skye usually heads straight upstairs to bed, I'm not surprised that only Charlie is waiting by the larder for his customary praise and ear rub, and the prospect of a tasty treat. Those given, I go into the living room and begin switching off the lamps.

A short, sharp bark calls me back to the kitchen. Charlie is standing by the closed door, his expression anxious. He looks pointedly at me, aims his nose at the door, and gives another cock-crowing bark. I open the door, thinking that he hasn't yet completed his evening business, but he steps back as a very relieved Skye runs inside, nuzzles Charlie gratefully, and flashes a reproachful look in my direction on his way past. Charlie understood that Skye had slipped back out without me seeing him, and knew just what to do in order to alert me that his friend needed to be let back in.

Watching a fearful dog gain confidence is an incredibly rewarding experience. This takes time and patience to accomplish, and it's necessary to first of all create a solid foundation of trust so that the dog feels safe enough to gradually expand his comfort zone. Hand-feeding Charlie when he first arrived and found the food and water bowls scary was the first step in gaining his trust. This was then built on through being consistently kind and gentle with him, to teach him that he could feel safe in my presence.

... how magnificent he must have been in the wild

Seeing the transformation in Charlie when his confidence blossomed made me think about how magnificent he must have been in the wild, where he understood his environment and could adapt to its changes through the seasons of the early years of his life. At heart, Charlie is a proud individual. Now that he feels comfortable and secure, he holds his head and tail high, and carries himself with loose, gliding movements that make him appear taller, as he seemingly floats just above the ground. His journey with me has gradually uncovered elements of his essential self, but Skye has played a considerable role in boosting Charlie's confidence and well-being.

Like us, dogs learn a great deal through observation and imitation, and Skye is an ideal role model, relaxed and at ease with people and other dogs. He embraces new experiences with shining eyes and an alert, interested attitude. Because he has always been treated well, he expects the world to be a good place. And he bonded strongly with Charlie through play during the early months.

THE PURPOSE AND POWER OF PLAY

Dogs play for a number of reasons. Play instigates and strengthens social bonds, creates connections between the neurons in the cerebral cortex of the brain, improves cognitive skills, allows predatory behaviours to be learned and honed in a safe space, and teaches cooperation and trust. There are distinct rules of engagement: dogs make an effort to play fair, because any infringement on the rules brings an immediate end to the game as the offended party walks away.

With Shep, our fifteen-year-old, terminally ill foster dog, Skye was careful to self-handicap in order to make sure that Shep wouldn't lose his fragile balance. He lowered his body and muzzle-fenced, instead of engaging in bumping or nudging. He also did this with Orla, a beautiful, ten-year-old ex-racing Greyhound I adopted when Skye was around a year old.

Orla and her two daughters were classed as three of the worst cruelty cases to be brought to the UK from the rescue in Ireland which took them in. They had been shut away in a dark shed for many months; starved, and doused in sheep dip chemicals. As a skeletal senior dog whose soft, white fur was pink from the chemicals that had burned her skin, Orla was considered harder to home. Each week my friend, Annie Rawlings, and I used to visit the rescue kennels, which, at that time, were run by our friend Annalisa (Ali) Cook, to walk the dogs, assess them, and make notes, and Annie often micro-chipped new arrivals.

We met Orla and her daughters the day after they arrived. They were in a pitiful state, and I lay awake all night thinking about them. The next day I rang Ali and said I wanted to adopt one of them. She suggested Orla, and I immediately agreed. Two days later Orla was brought to us by Carol, a colleague. We walked Skye and Orla together, they made it clear they liked each other, and we headed home.

Carol and I both had tears in our eyes, seeing the pure joy in the face of a dog who had never had a home

---

Orla stood beside me as I unlocked the front door. She looked at the door, looked at me, and her face lit up. It was as if she knew she was home, and Carol and I both had tears in our eyes, seeing the pure joy in the face of a dog who had never had a home, had been so cruelly treated, and yet was still determined to enjoy life to the full. She settled in wonderfully, but two days after she arrived I found a mammary tumour when she lay on her back on the sofa (it's called 'roaching' when Greyhounds do this), inviting me to stroke her belly. I took her straight to the vet, who operated on her a few days later, and she had nine very happy months of frolicking with Skye, making up for all the love she'd never received in the past. Her coat shed and grew back, she gained weight, and was such a sweet-natured dog that a number of people adopted Greyhounds after meeting her.

During the tenth month Orla became very ill. She lost weight rapidly, her fur fell out, she bumped into things, fell over a lot, seemed vague and confused, and pressed her head hard against me at every opportunity. She became phobic about the smooth kitchen floor, as she couldn't negotiate it easily. The vet checked her over, told me that our beautiful girl had lost her sight and hearing, and diagnosed a brain tumour. All we could do was keep her comfortable and pain-free during her final month.

Yet this brave, beautiful soul was courageous to the end, even when she began to have seizures, and we knew the kindest thing was to let her go in peace. Up until her last few days she still had moments of playing with Skye – Orla lying on the sofa; Skye standing on the floor in front of her, gently muzzle-fencing. She took a piece of my heart with her when she left us.

Charlie is a different character altogether. He plays rough, and when Skye's health was more robust he tolerated being pounced on, jumped on, and body-slammed, and only expressed disapproval if Charlie became over-excited and curled his toes to dig his claws into Skye's back. A side-step and low rumble made the message clear. Charlie would immediately play-bow to signal that he meant no harm and wanted the game to continue, and Skye forgave him the transgression, bowed in return and leaped forward for the next round. Play is gentler now that Skye is frail, and it is Charlie who makes an effort to self-handicap by muzzle-fencing and curbing his tendency towards hard body-slams.

He loves to lie out on the lawn
on sunny days

During the first few months, the boys played throughout most of their waking hours. This continually reinforced their bond, and was instrumental in building Charlie's confidence around us, as well as with Skye. It felt like a huge step forward in our relationship when Charlie began to invite me to play.

He loves to lie out on the lawn on sunny days, absorbing the warm rays and taking in the scents and sounds. When I step out into the garden he immediately jumps up and runs over, play-bows, then dances around me while I neatly side-step to avoid his sharp extended claws before bending to ruffle his fur. He weaves in and out of my legs, curves his body to rub against me, head-butts, nudges and stands on his hind legs to place his paws in my hands for a brief waltz. It has taken a long time for this game to develop, and Charlie always makes it abundantly clear that his joy in it matches my own.

It's likely that our relationship would have been very different if not for my background in working with troubled dogs. I needed to feel confident about my ability to find the most effective ways in which to gain Charlie's trust and affection. It was vital to be able to understand the signals Charlie was giving, and to tailor my body language to reassure him, and each dog I have worked with in the past has contributed toward the skills I utilised for helping Charlie. Good relationships are built on the bedrock of mutual trust and respect, and are further enhanced through creating the capacity to simply have fun together; to enjoy each other's company.

THE FEEL-GOOD FACTOR

Another key to developing confidence is positive feedback and encouragement. This is something that we all need and appreciate, and acknowledgement makes a huge difference to how we feel, regardless of which species we belong to. The Sympatico Method that I wrote about extensively in my book *The Heartbeat at Your Feet* has a strong focus on positive feedback, and on rewarding the behaviours that we wish to encourage. An action or approach that ultimately results in us feeling good about ourselves is more likely to be repeated in the future, and this worked very effectively with Charlie.

Our feral boy had a great many challenges to contend with

Our feral boy had a great many challenges to contend with. Experiences that our home-reared dogs easily accept were unfamiliar, and therefore very frightening for him. To be constantly confronted by what he perceived as potential danger in every moment of his early days must have been hugely stressful for Charlie, and a step-by-step approach that involved setting him up for success during each new experience enabled Charlie to gradually overcome his fears and gain confidence.

When we are encouraged as we progress; when we know that someone believes in us and is rooting for us, this gives us the confidence to keep going – to push through barriers created by fear and anxiety. Each small achievement becomes a stepping stone to greater achievement. It's the same with dogs, and Charlie was no exception. We had setbacks when Charlie reverted to his previous nervous self, but each of these reduced in duration as the months passed, and Charlie soon returned to whatever level he had reached before fear stopped him temporarily in his tracks. I learned to expect this, and often to anticipate it. An influx of visitors, someone approaching him too quickly for comfort, or a day where I felt emotional were all triggers for Charlie's regressions.

When Amber went away on holiday for five days, Charlie became very anxious. He reverted to skittering around, his belly close to

the ground. His appetite diminished, and Skye took advantage of the ignored food by helping himself to an extra snack when it was clear that Charlie wasn't interested in eating. Several times I watched in amusement as he sidled in slow motion towards the bowl, pretending he hadn't noticed me in the room, then looked at me in apparent surprise when I said, "Skye, you are so busted, my friend!" as he dipped his muzzle to eat.

Skye did his best to cheer up Charlie by inviting him to play, and Pete, one of our neighbours, came out for walks with us with his Springer Spaniel, Mayzie, who Charlie had become very fond of. Back at home, he moped. His head and tail were carried low, he slunk around, and he stopped instigating contact.

Amber arrived home around three in the morning. I'd waited up for her, eager to hear about her adventures. Both dogs were fast asleep, but as soon as they heard her key in the door they leapt up and ran to greet her in a frenzy of squeaks, whines and ecstatic wriggles. The next morning Charlie was back to normal, eagerly waffling down his breakfast, and play-bowing enthusiastically at Skye.

A year later, Amber went away for a week with her boyfriend, Sam. I expected some reaction from Charlie, but this time he coped beautifully and was completely relaxed throughout her absence. The pair arrived home in the middle of the night, to be greeted happily by two sleepy dogs.

Decisions and choices

Love involves mutual respect. I am the decision-maker in Charlie's life because he isn't equipped to understand the world he now inhabits. Charlie now respects that, just as I respect that he is a soul who has grown up used to making what were likely to be life or death decisions for himself. Although his freedom is limited in comparison to his previous life, we strike a balance because I allow Charlie to make his own choices – where he sleeps, the position of his food bowl (and he would rather go hungry than eat from a bowl that's in the wrong position), the route we take through the field and woods. We are a team, and this promotes harmony as well as increasing confidence.

Relationships are multi-faceted and work both ways. I've helped Charlie to gain confidence, even in unfamiliar situations, and he has proven conclusively that 'positive,' force-free methods are the best way in which to help even a wild creature adjust to a completely new way of life. Charlie has confirmed to me what I've always felt very deeply – that love and the right kind of care transcends all barriers and boundaries.

sixteen
Love conquers all (eventually)

Sixteen months after Charlie's arrival, I take him and Skye for a walk on my own. This is a recent development. Usually, Amber and I go out together with them, so that both dogs can move at their own pace. Skye has slowed down a great deal, and prefers to amble along, catching up on news that other dogs have left in the field, while Charlie likes to trot, and still has a habit of either bolting or freezing when he sees anything unfamiliar.

The boys dance with glee when I lift their harnesses and leashes from the peg in the hall, and stand quietly while I slip them over their heads and buckle them. Once out, they walk side by side, stopping to sniff around when some scent catches their attention. Charlie deliberately slows his pace so that Skye can keep up with him comfortably, and I'm touched that he's so considerate of his friend. We greet people as we pass them, and one of my neighbours, who Charlie has never met, stops for a chat. To our surprise and delight Charlie walks straight over to her, sits to offer both front paws, and invites her to stroke him.

I fell in love with Charlie the moment Gina carried him, trembling and paralysed with terror, into our home on that cold February evening in 2013. His fear and confusion brought out the protective instinct in me; the desire to help heal the psychological wounds, as has happened with all of the traumatised dogs I've worked with over the years. Love and nurturing, combined with helping Charlie to feel safe, allowed the first step to healing, and Charlie's behaviour leaves me in no doubt that he loves me too.

Charlie's behaviour leaves me in no doubt that he loves me too

---

Not so long ago it was considered bad science to attribute emotional states to non-human animals, despite Charles Darwin – the first scientist to closely study emotions in animals – maintaining that he had witnessed anger, happiness, sadness, disgust, fear, and surprise in the animals he observed. Darwin had a close relationship with his own dog, and wrote of his observations in *The Expression of the Emotions in Man and Animals*, which was published in 1872. In this book, which came out thirteen years after *On the Origin of Species*, Darwin explored the similarities between those facial expressions that register emotion in both humans and animals. Nowadays, scientists who study animal emotions and cognition have added to the list of emotions that are clearly expressed by non-human animals, and are convinced that the range is very close to our own emotional spectrum.

The visible, tangible bond that developed between Charlie and I prompted me to think about the nature of love. This emotion is at the heart of all significant relationships, whether these are familial, romantic, or the close connection between friends. Its essence is the desire to be with another, to wish for their happiness, and to express feelings of deep warmth and affection. The chemistry of love is physical, prompted by the release of hormones which have a profound affect on how we feel.

THE CHEMISTRY OF LOVE

First of all come the neurotransmitters: adrenaline, dopamine, and serotonin. These provide the jolt of recognition, the accelerated heart rate, the intense rush of pleasure in the presence of the loved one. Then there's oxytocin, the 'love hormone' that is released during the bonding process between a mother and her baby. It's been proven that when we stroke our dogs there's a rise in oxytocin levels in the bloodstreams of both parties. Endorphins, which stimulate feelings of attachment and comfort, carry the relationship through in the long-term. In his beautiful book *The Emotional Lives of Animals,* Marc Bekoff points out that there is plenty of evidence to show that non-human animals are able to feel the various forms of love (filial, parental, erotic, romantic, and platonic), just as we do.

Relationships may begin with a flash of recognition and connection, or may blossom slowly, but they take time to fully develop, and both parties need to invest energy in them. Charlie and I have grown together. We've learned a great deal from each other, and the foundation we have built between us will always be there as his range of new experiences broadens. We've grown extraordinarily close through the many highs and lows during the first eighteen months together, and Charlie's strong bond with me, Amber, and Skye has helped him to adapt to the huge changes in his life. Between us we have

worked through his many fears and anxieties, and overcome serious behaviour issues. This process was mutually cooperative, and could only take place because Charlie trusted me, and was willing to travel this path to healing with me.

We've grown extraordinarily close through the many highs and lows

The bedrock of our relationship is respect, as well as love: the two go hand-in-paw. I respect Charlie's wild nature, his courageous heart, his instinctive responses, his wolfishness and doggyness; the hilariously crazy puppyish side that's emerged, and his right to make choices about who he wishes to interact with. Charlie respects my calm, gentle guidance, and my determination to keep him safe while he explores his new world and forms new friendships.

The relationship between Charlie and I could not be categorised – as it often is with our domestic dogs – as similar to a parent-child bond. We are friends and partners, guiding each other through the labyrinth one step at a time, enjoying the scenery along the way, even if sometimes the emotional weather includes some storms. Despite Charlie's fears of the new and unfamiliar, he is not a meek, mild, malleable soul, but strong and fierce and proud; a creature who, for his first few years, was autonomous, in complete control of his life. With Skye, Charlie is the younger brother who looks up to his mentor, learns from him, and instigates fun. His relationship with Amber is remarkably similar to the one he has with Skye. When Charlie is worried, however, he comes straight to me, and the trust he now has in me makes my heart sing.

Of course, there may be obstacles and challenges for us to negotiate through in the years ahead: that is the nature of life, and of relationships. Yet, we have come such a long way, my feral friend and I, and the bond between us is immensely strong. We have all that is needed to see us through even the most difficult patches. To journey along life's pathways with any creature is a blessing, and the gift of Charlie's presence is a privilege I hadn't anticipated and am very grateful for.

The emergence of nurturing and acceptance

A while ago, Skye had another severe bout of the colitis that has been ongoing since his long illness the previous year. It was upsetting to see him so unwell, and I couldn't help but shed some tears. Up until that day, Charlie's reaction to emotional upheaval had been aggression, and his response to Skye's frailty and my emotional upset revealed just how far he has come and how deeply he loves us. He anxiously watched over Skye, lying close by to lend support, and was especially sweet and gentle in his interactions with me.

That night he put both front paws on my bed and looked askance at me. I told him he was welcome to join me, and he sprang up and lay down at the foot of the bed, rested his chin on my legs and slipped into a deep, peaceful sleep. Since then he has joined me each night, respectfully asking permission first of all before leaping up to his favourite spot at the end of the bed, making sure that some part of him finds a resting place against me.

This recent tolerance of vulnerability in Charlie has extended beyond our immediate family. His eventual acceptance of Sam, Amber's boyfriend, towards whom he had displayed a marked antipathy for such a long time, was a relief and a delight for all of us. No-one, whatever the species, should be put in a position of having to interact when they don't wish to, and Sam's willingness to (very sensibly) give Charlie plenty of space, combined with the instigation of an association between Sam's arrival and a tasty reward paired with those oh-so-welcome ear rubs finally bore fruit.

Charlie has brought many lessons my way, and I've absorbed these gladly and with gratitude. I feel privileged to live with him. Being with Charlie – my feral boy, my wild soul – has shown me that we each retain the wild within us, and that nature is not always red in tooth and claw. It has its own rules, its methods of justice and reparation, and the instinctive responses that we have in common with other mammals include love, empathy and compassion.

This morning, I opened my eyes as the early rays of sunshine filtered through the curtains. Charlie was stretched out alongside my legs, a picture of relaxation and security, with Skye asleep on the floor, close beside my bed. As I lay on my side, watching Charlie and savouring the warm glow that accompanies the presence of loved ones, he felt my gaze, woke, and raised his head. His golden-amber eye met mine and we shared a long, tender look. After a while, he wriggled up the bed to rest the side of his head against my hand in a gesture that told me clearly that he is as happy to be with me as I am to be with him.

Love has no boundaries and knows no scarcity or lack. The more we open our hearts, the more our capacity to love expands. Love is the healer, the life-enhancer, the bright light that shines from within, illuminating and driving away shadows, transforming us and those around us who allow themselves to give and to receive from its bounty. Charlie has taught me so much about unconditional love and the joy that this brings. He has taught me to listen to the voice of the wild self that dwells deep within, to pay attention to that quiet whisper and rejoice in its presence.

With such a friend and teacher, love is the stepping stone from one bright moment to another; a priceless gift.

VISIT HUBBLE AND HATTIE ON THE WEB: WWW.HUBBLEANDHATTIE.COM
WWW.HUBBLEANDHATTIE.BLOGSPOT.CO.UK
• DETAILS OF ALL BOOKS • SPECIAL OFFERS • NEWSLETTER • NEW BOOK NEWS

seventeen POSTSCRIPT

Charlie came to live with us in February 2013, and I delivered the manuscript for this book to Jude at Hubble & Hattie in October 2014, twenty months after Charlie's arrival, so the chapters describe our adventures together up to that point in time. Charlie's progress continued after that: he made many new friends, and became a kind of canine figurehead for rescue dogs. The bond of love and trust between us kept on growing, and sharing updates and photos of our sweet boy brought him a huge following of friends on Facebook.

In February 2015, Charlie became very unwell. He rallied after a few days but his energy level was lower than before. He began to lick himself compulsively, and in late March he started pressing his head hard against me. This worried me as head-pressing can be a red flag for liver or brain issues. Blood tests flagged up higher than normal liver values, so MRI scans were booked for April 27, when Alex Gough, the specialist referral vet, returned from holiday. In the meantime, Amelia prescribed Metacam and, at her suggestion, I also gave Charlie herbal Skullcap and Valerian, plus a nutraceutical called Zylkene, to help him remain calm.

Despite his health problems, Charlie was happy most of the time. Friends visited and he charmed them all. He even bonded with Stef and Leon, my friends who have wolfdogs – the friends he'd refused to interact with the previous year. His relationship with Amber's boyfriend, Sam, blossomed to best-buddy proportions, and he would dash to the door to greet him as soon as he heard his voice. He slept on my bed, tucked close against me, and spent the evenings curled up beside me on the sofa. He played with Skye, though less than before. Amelia monitored him closely but there was frustratingly little that we could do without further investigations. We counted down the days to Alex's return and the scan appointment.

The week before Charlie's scans were due his health deteriorated frighteningly fast. He refused food and spent most of the time on my bed, sleeping. His sad "Please help me" expression was heartbreaking to see, and he avoided touch. On Friday, April 24, Amelia visited and expressed deep concern about whether we would be able to pull him through. She told me to call her out over the weekend if he slipped further down. All we could do was hope that he would rally enough to have the scans in three days, and that these would show us a way forward.

On Sunday Charlie perked up. Our friend, Annie, one of Charlie's favourite people, came round and gave him Reiki healing, which he'd responded well to during the previous weeks. Charlie was his old exuberant self for a couple of hours. He smooched with Annie, offered paws, rested his head on her hand, and blatantly invited ear rubs. He snuggled against me on the sofa, and then ate a small meal after Annie left. We were thrilled and once again hopeful.

My son, Oliver, called round later for a while. He was another favourite friend, and Charlie greeted him cheerfully, though by then his energy was flagging. What turned out to be Charlie's final full day with us was infused with joy: his, and ours. That night he slept with his head on my legs while I lay awake, savouring his warm presence and worrying about what the next day would bring.

The following morning Charlie had very bloody stools. Amelia arrived just after 10am, and I told her about this. As we'd be in the dark without a full diagnosis, we decided the only real option was to go ahead with the scans. Then Charlie collapsed and stopped breathing.

Amelia swiftly put him in the recovery position, telling him to breathe, breathe, breathe, and he started breathing again, though shallowly. A mad dash to the hospital ensued. Amelia ran out to her ambulance and threw open the doors as Amber carried Charlie out. We made him comfortable and Amelia drove off at speed, with us following in the car.

Once at the hospital Alex took over, but Charlie didn't wake up. Amelia stayed with him throughout, and Alex called me while we waited nearby to tell me that nothing more could be done. He came out to the reception area and Amber and I followed him through to the scanning room. Charlie's breathing tube was removed and the room was cleared of all the medical staff except for Amber, Amelia, Alex and myself. We stroked and kissed Charlie, said goodbye to him, and thanked him for every precious moment that he'd given us. We all cried.

It felt to me afterwards that, really, Charlie had flown free at home, in our living room. He had no awareness of being at the hospital, his passing was peaceful, with no pain or fear, and for that, at least, I am grateful.

As I write this postscript, only two days have passed since Charlie left us. We miss him. He was a huge presence in our lives and our home, and the space he left behind is vast. Time will heal the rawness of sorrow and leave us with the best of memories, but that is yet to come. Charlie was our beloved friend, a larger-than-life character, a free spirit, a mischief maker, a hole digger, a cuddler and snuggler, a singer: a dog who made the most of every opportunity for joy – and who taught me, and others, to do that, too.

When I shared the devastating news of Charlie's death with my friends on Facebook there was a vast outpouring of grief from people from all around the world who had followed his story for over two years. Many had never met him in person but had been deeply touched by him. I realised that Charlie wasn't just our dog; he was everyone's dog. He had captured hearts and raised awareness of the sad plight of many other dogs who needed homes. Charlie changed many, many lives. Thousands of messages and comments told of the love that people felt for him, the tears they were shedding along with us, and stories of how he had inspired them.

Here's an extract from the tribute I wrote for Charlie and shared on Facebook on our first evening without him, when it was hard to imagine that he wouldn't come bouncing downstairs or in from the garden.

Charlie was the most extraordinary being I have ever known, and the void left by his passing is immense. The purity of his wild soul shone through him like a guiding light, and he taught me so much about how we humans seek to find ways to return to this natural state. His trust and love were among the greatest gifts I have ever received, and I treasure every moment we shared together.

Charlie

What a journey we travelled, side by side. We faced many challenges, built bridges to mutual understanding, and met, hand-in-paw, in the middle. We learned to match our steps and find a pace that was comfortable for us both while we negotiated rocky passes and deep water, and we gladly succumbed to the bone-melting, heart-melding experience of unconditional love.

And there, all the time, was Skye. Friend and mentor, teacher and wise guide, whose gentle presence gave Charlie the confidence to trust that he could feel safe in the scary new world he had been thrust into. There was such a bond between these two very different, very beautiful souls.

Charlie taught me so much. He reminded me of the wisdom that is gained through stepping back and looking through fresh eyes to take a new perspective. He made me laugh with his clownish antics, wild dances, and full-throated howling sessions. He was fierce and strong, ferociously loyal, generously affectionate, and he wore courage like a badge of honour along with his scars from past combats.

I'm deeply grateful for the time we had together. It was special because Charlie was special. There will never be another like him, and the sense of loss that Amber and I feel goes beyond mere words.

What I will choose to hold on to is the love. What will stay forever in my heart is the luminous memory of the emergence of trust and love between a wild soul and the humans in his life, and the bond between a feral dog with a past we can hardly even imagine and a Lurcher who helped to guide him through the terrors of domestic life towards acceptance and joy.

So many people who met Charlie fell in love with him. Many more who never had an opportunity to meet him also loved him. A lot of tears have been shed today in our home and in other homes around the world. We were very blessed to know him.

Run free, my love. Run fleet-footed through light-filled fields and woods with your tongue lolling in that wonderful smile. Dig holes, and make them big ones. Run free of pain and suffering, in a place where all that exists is love. You'll live always in our hearts, and when the time comes I will run to you and with you.

Charlie was a bridge who connected hearts all across the world with his own huge, pure and honest heart. He slipped so deeply into my heart that he became part of the fabric of my very soul. His time with us was too brief, but his memory will be forever treasured and honoured.

Thank you for sharing Charlie with us through reading his story. Perhaps he was meant to come to us for just long enough to teach us about the wild self, and to learn what it is to be loved and cherished. I wished for years with him, but I'm holding an image of Charlie dancing around in a place of light and freedom, healed and whole, howling at the moon, digging holes ... and smiling his beautiful smile.

ABOUT THE AUTHOR

Lisa is the founder and principal of The International School for Canine Psychology & Behaviour, a globally approved college that runs courses for students from all around the world. She is the founder of the Dog Welfare Alliance, a non-profit organisation which brings together members of the public and dog professionals, and supports rescue organisations across the globe. She is also Chair of the Association of INTO Dogs, an organisation for professionals who use only 'positive,' force-free methods. Lisa has fostered and adopted many rescue dogs, and she is passionate about dog welfare.

Lisa is the author of twenty-six published books, including three previous books about dog psychology and behaviour. She is a consultant for the BBC on dog-related issues.

InDex